Green Chromatographic Techniques

Inamuddin • Ali Mohammad
Editors

Green Chromatographic Techniques

Separation and Purification of Organic and Inorganic Analytes

Springer

Editors
Inamuddin
Department of Applied Chemistry,
Faculty of Engineering & Technology,
Aligarh Muslim University,
Aligarh
India

Ali Mohammad
Department of Applied Chemistry,
Faculty of Engineering & Technology,
Aligarh Muslim University,
Aligarh
India

ISBN 978-94-007-7734-7 ISBN 978-94-007-7735-4 (eBook)
DOI 10.1007/978-94-007-7735-4
Springer Dordrecht Heidelberg London New York

Library of Congress Control Number: 2013952774

Printed on acid-free paper

Springer is part of Springer Science+Business Media (www.springer.com)

Preface

Green technologies demand the use of safer instrumental techniques for the separation, identification and purification of organic and inorganic analyte. Nowadays, special emphasis has been given to the separation and purification of bioactive components like vitamins, amino acids, toxins, carbohydrate etc. However, the concern about the use of toxic solvents has been completely ignored. Therefore, to make the system convenient, it is advisable to use non-toxic solvents or techniques for the separation and purification. So that life can sustain a long without exposure of harmful toxic solvents.

This book would be a wide spectrum scientific resource on the use of green chromatographic techniques for the separation, identification and purification of bioactive as well as inorganic analyte. The aim of this book is to provide in depth knowledge of the green techniques which utilize green solvents. In some chapters the technique itself is green or using non toxic environment. However, the book is basically aimed to merge the terms green techniques and green solvents as **green chromatography** which may prove the useful resource to the scientist working in the field of analytical, organic and pharmaceutical chemistry.

Green Chromatographic Techniques: Separation and Purification of Organic and Inorganic Analytes' edition with *most up-to-date reference* work will prove a necessary resource for scientists, R&D industrial specialists, researchers, upper-level undergraduate and graduate students, Ph.D. students, college and university professors working in chemistry, chemical and biochemical fields. Based on thematic topics, the book edition contains the following 9 chapters:

Chapter 1 addresses some practical and theoretical aspects of counter current chromatography (CCC), highlighting the specific advantages of this support-free liquid stationary phase purification greener technique.

Chapter 2 is dealing with new preparative method of separating concentrated solutions of mineral electrolytes into individual components by size exclusion chromatography on neutral nanoporous hyper-crosslinked polystyrene sorbents. Basic principles of the method as well as factors determining the selectivity of separations are discussed.

Chapter 3 is dealing with the supercritical fluid chromatography which is known as a green approach for the separation and purification of organic and inorganic analyte.

Chapter 4 provides historical development of thin-layer chromatography towards becoming modern, automated, high resolution technique in the form of high-performance thin-layer chromatography, and their further advances in miniaturization of chromatographic beds in the form of ultra-performance thin-layer chromatography (UPTLC).

Chapter 5 highlights the different aspects of gas chromatography in the light of green techniques starting from sample preparation to the selection of mobile phase and chromatographic columns. Coupling other analytical tools with GC to focus the versatility and high accuracy of analysis with dual system of separation and detection is also discussed.

Chapter 6 is dealing with the green reversed-phase high-performance liquid chromatography (RP-HPLC), and green thin layer chromatography (TLC) methods used for preparing and purifying allicin, a garlic-derived organosulfur compound.

Chapter 7 is dealing with some important green sample preparation techniques used for the separation of organic analyte in complex matrices.

Chapter 8 is focusing the determination of organochlorine pesticides was made through the gas chromatography method using capillary columns and detector with electrons capture.

Chapter 9 describes the retention mechanisms for size exclusion chromatography (SEC) and applications of SEC in the biomedical and pharmaceutical sciences. Finally, the use of SEC as a technique for speciation analysis of polydimethylsiloxanes (PDMS) is presented.

Inamuddin
Ali Mohammad

Acknowledgements

We are most indebted to the grace of the Almighty "One Universal Being," who inspires the entire Humanity to knowledge and who has given us the required favor to complete this work.

This book is the outcome of the remarkable contribution of experts from various interdisciplinary fields of science and cover the most comprehensive, in-depth, and up-to-date research and reviews. We are thankful to all the contributing authors and their coauthors for their esteemed work. We would also like to thank all publishers, authors, and others who granted us permission to use their figures, tables, and schemes.

We would like to express our deep gratitude to Prof. T. Urushadze (Georgia State Agriculture University, Georgia), Prof. K. Aoki (Toyohashi University of Technology, Japan), Prof. Rajeev Jain (Jiwaji University, India), Prof. S. Shtykov (Saratov State University, Russia), Prof. M. M. Srivastava (Dayal Bagh University, India), Prof. M. C. Chattopadhyaya (Allahabad University, India), Prof. U. S. Roy (Visva-Bharti Santiniketan, India), Dr. Ajay Taneja (Dr. B. R. Ambedkar University, Agra, India), Prof. A. P. Gupta (Delhi Technological University, India), Prof. Anca Sipos (Lucian Blaga University of Sibiu, Romania), Prof. J. K. Rozylo (Maria Curie-Skloclowska, Poland), Prof. P. K. Sharma (JNV University, Jodhpur), Prof. Ravi Bhusan (I.I.T. Roorkee, India), Prof. Ibraheem (Jamia Millia Islamia, India), Prof. El-Sayed Ali Abdel-Aal (CMRDI, Cairo, Egypt), Dr. Ajay Taneja (Dr. B. R. Ambedkar University, India), Dr. Reeta Mehra (MDS University, Ajmer), Prof. M. Kidwai (University of Delhi, India), Prof. M. S. Chauhan (Himachal Pradesh University, India), Prof. A. S. Aswar (SGB Amaravati University, India), Dr. Anees Ahmad and Prof. Syed Ashfaq Nabi (Department of Chemistry, Aligarh Muslim University (A.M.U.), India), Prof. J. Sherma (USA) and Prof. M. Mascini (University of Firenze, Italy), Prof. Ishtiaq Ahmad and Prof. Rakesh Kumar Mahajan (Department of Chemistry, Guru Nanak Dev University, Amritsar, India), Dr. B. D. Malhotra (Scientist-F, NPL, New Delhi, India), Dr. Raju Khan (Scientist-C, NEIST, Assam, India), Prof. Seon Jeon Kim (Hanyang University, South Korea), Prof. Kenneth I. Ozoemena (University of Pretoria, South Africa), Prof. Saleem-ur-Rahman and Prof. S. M. J. Zaidi (King Fahd University of Petroleum and Minerals, Saudi Arabia), Prof. Gaber Eldesoky and Prof. Zeid-AL-Othman (King Saud University,

Saudi Arabia), Prof. Sheikh Raisuddin (Jamia Hamdard University, New Delhi, India), Byong-Hun Jeon (Yonsei University, South Korea), and Prof. A. I. Yahya (Nizwa University, Oman) for their valuable suggestions, guidance, and constant inspiration.

It is with immense gratitude that we thank our departmental colleagues Prof. M. Mobin, Prof. Asif Ali Khan, Prof. R. A. K. Rao, Prof. Faiz Mohammad, Dr. M. Z. A. Rafiquee, Dr. Abu Nasar, Dr. Rais Ahmad, and Dr. Yasser Azim, without whose continuous encouragement these books would have not been completed. Dr. Inamuddin cannot thank enough his friends and colleagues Dr. M. M. Alam (USA), Dr. Amir-Al-Ahmad (KFUPM, Saudi Arabia), Dr. Zafar Alam, Dr. Mu. Naushad, Dr. Mohammad Luqman, Dr. Salabh Jain, Dr. Hemendra Kumar Tiwari, Dr. Adesh Bhadana, Dr. Shakeel Ahmad Khan, Satish Singh, and others, for their timely help, good wishes, encouragement, and affections.

The help received from our research group (Dr. Arshi Amin, Tauseef Ahmad Rangreez, Ajhar Khan, Khursheed Ahmad and Sardar Hussain) is appreciatively acknowledged.

Finally, we feel short of words and full of emotions in thanking our family members for their constant inspiration and gracious support.

Inamuddin
Ali Mohammad

Contents

Contributors

Filip Andrić Faculty of Chemistry, University of Belgrade, Studentski trg 12-16, 11000 Belgrade, Serbia

Runa Ghosh Auddy Department of Chemical Technology, University of Calcutta, 92 A. P. C Road, Kolkata 700009 West Bengal, India
e-mail: auddyruna@gmail.com

Alain Berthod Institut des Sciences Analytiques, CNRS ISA UMR 5280, Université de Lyon, 5 rue de la Doua, 69100 Villeurbanne, France

Dipa Biswas Department of Chemical Technology, 92, A P C Road, 700009, Kolkata, India
e-mail: dipa_bis@yahoo.com

Zinaida Blinnikova Nesmeyanov-Institute of Organo-Element Compounds, Russian Academy of Sciences, 119991 Moscow, Russia

Adriana Chiş Department of Food control, Environment Protection Faculty, University of Oradea, Oradea, Romania
e-mail: andichis@yahoo.com

Purcărea Cornelia Department of Food control, Environment Protection Faculty, University of Oradea, Oradea, Romania

Sriparna Datta Department of Chemical Technology, University of Calcutta, 92 A. P. C Road, Kolkata 700009 West Bengal, India
e-mail: sriparna_d@yahoo.com

Vadim Davankov Nesmeyanov-Institute of Organo-Element Compounds, Russian Academy of Sciences, 119991 Moscow, Russia

Amit De Department of Chemical Technology, University of Calcutta, 92 A. P. C Road, Kolkata 700009 West Bengal, India

Karine Faure Institut des Sciences Analytiques, CNRS ISA UMR 5280, Université de Lyon, 5 rue de la Doua, 69100 Villeurbanne, France
e-mail: karine.faure@univ-lyon1.fr

Shalini Gupta Protein Research Laboratory, Research Resources Center, University of Illinois at Chicago, 835 S. Wolcott Avenue, Chicago, IL 60612, USA

Cristina Horga Institute of Public Health, Cluj-Napoca, Romania

Jin-Sheng Huang Protein Research Laboratory, Research Resources Center, University of Illinois at Chicago, 835 S. Wolcott Avenue, Chicago, IL 60612, USA

Lasanthi P. Jayathilaka Protein Research Laboratory, Research Resources Center, University of Illinois at Chicago, 835 S. Wolcott Avenue, Chicago, IL 60612, USA

Bao-Shiang Lee Protein Research Laboratory, Research Resources Center, University of Illinois at Chicago, 835 S. Wolcott Avenue, Chicago, IL 60612, USA
e-mail: boblee@uic.edu

Jenny Lee Illinois Mathematics and Science Academy, 1500 Sullivan Rd, Aurora, IL 60506-1000, USA

Nazim Mekaoui Institut des Sciences Analytiques, CNRS ISA UMR 5280, Université de Lyon, 5 rue de la Doua, 69100 Villeurbanne, France
e-mail: davank@ineos.ac.ru

Dušanka Milojković-Opsenica Faculty of Chemistry, University of Belgrade, Studentski trg 12-16, 11000 Belgrade, Serbia
e-mail: dusankam@chem.bg.ac.rs

Debarati Mitra Department of Chemical Technology, 92, A P C Road, 700009 Kolkata, India
e-mail: debarati_che@yahoo.com

Krystyna Mojsiewicz-Pieńkowska Faculty of Pharmacy, Department of Physical Chemistry, Medical University of Gdańsk, Al. Gen. Hallera 107 Gdańsk 80-416, Poland
e-mail: kpienk@gumed.edu.pl

Cíntia Alessandra Matiucci Pereira Centro Universitário Central Paulista (UNICEP), R. Miguel Petroni 5111, São Carlos, SP 13563-470 Brazil
e-mail: cmatiucci@gmail.com

Maria Tsyurupa Nesmeyanov-Institute of Organo-Element Compounds, Russian Academy of Sciences, 119991 Moscow, Russia

Vânia Gomes Zuin Universidade Federal de São Carlos (UFSCar), Rodovia Washington Luiz, Km 235, São Carlos, SP 13565-905 Brazil
e-mail: vaniaz@ufscar.br

Chapter 1
Saving Solvents in Chromatographic Purifications: The Counter-Current Chromatography Technique

Karine Faure, Nazim Mekaoui and Alain Berthod

Abstract This chapter addresses some practical and theoretical aspects of counter current chromatography (CCC), highlighting the specific advantages of this support-free liquid stationary phase purification technique. The focus is on the latest instrumental developments which demonstrates that the poorly known CCC technique may exhibit a high, unrealised potential for greener purification processes.

1.1 Introduction

Counter-current chromatography (CCC) is a purification technique that is based on the partitioning of the solutes between two immiscible liquid phases, one being immobile (i.e. stationary phase), while the other liquid (i.e. mobile phase) passes through. This definition includes both the advantages and drawbacks of the technique. First to be noticed, there is absolutely no mention of any counter current flow. Indeed, only one liquid phase flows through the CCC instrument. The unfortunate choice of the technique name originates from historical reasons (Conway 2011) and will prevail as the cute abbreviation "CCC" became so popular. Secondly, the separation of solutes is based only on their distribution between two liquid phases. This is the simplest mechanism that can be found in chromatographic separations since one liquid phase is the mobile phase and the other liquid phase is the stationary phase. The characteristic feature of total absence of solid matrix in the CCC instrument promises a huge versatility for purification of complex samples.

If CCC does not ring any bell to the reader, the main reason is that solid/liquid chromatography has long established itself as the main purification technique, with a large instrument development over decades, while CCC has long suffered from poor engineering and unreliable instruments.

K. Faure (✉) · N. Mekaoui · A. Berthod
Institut des Sciences Analytiques, CNRS ISA UMR 5280,
Université de Lyon, 5 rue de la Doua, 69100 Villeurbanne, France
e-mail: karine.faure@univ-lyon1.fr

Inamuddin, A. Mohammad (eds.), *Green Chromatographic Techniques*,
DOI 10.1007/978-94-007-7735-4_1, © Springer Science+Business Media Dordrecht 2014

This review will address the theoretical aspects of CCC, highlighting the specific advantages of the technique. It will focus on the latest instrumental developments and demonstrate that the poorly known CCC technique may exhibit a high, unrealised potential for greener purification processes.

1.2 CCC Theory

The CCC retention mechanism is based on solute partitioning between the two non miscible liquid phases. The solute retention relies only on its distribution ratio, K_D, between the two phases. No adsorption occurs, as no solid material is present. The theory behind the separation mechanism is expressed by the simple equation:

$$V_R = V_M + K_D \cdot V_s \quad (1.1)$$

where

V_R is the retention volume,

V_M and V_S are, respectively, the mobile and stationary phase volumes inside the CCC instrument of total volume, V_C (=V_M+V_S),

K_D is the solute distribution ratio between the two phases that is the concentration of all forms of the solute in the stationary phase divided by the concentration of all forms of the solute in the mobile phase (Berthod 2002). K_D is also called solute liquid-liquid partition coefficient.

1.2.1 High Loadability

Because the separation process occurs solely in liquids allowing for a volume access, the sample mass that can be injected and treated in one run is generally more important in CCC than in conventional preparative liquid chromatography (prep-LC). In prep-LC, the stationary phase is a porous solid, most often silica particles. The solutes interact with the limited surface of the solid stationary phase that is rapidly overloaded. In CCC, the solutes interact with the much larger volume of liquid stationary phase so the amount of sample that can be introduced in the CCC column is much larger than in prep-LC, only limited by solubility. The samples can be solubilised either in the mobile phase or in the stationary phase, or even in a mixture of both phases.

The useable volume of liquid stationary phase can make up to 70–80 % of the overall CCC column volume, so overloading happens at significantly higher loadings compared to prep-LC where the useable solid stationary phase volume makes rarely more than 1 % of the column volume. The capacity of purification of the CCC technique is therefore significantly enhanced compared to classical prep-LC working with similar volume columns. Consequently, the number of injections/runs to produce a certain amount of pure compound is reduced. High loadability is the major incentive that CCC purification solution has. This appealing potential throughput can provide a fair solution to purify otherwise unprofitable products.

1.2.2 Scale up Capability

A CCC separation is governed solely by the distribution of the sample between the two immiscible liquid phases; hence its scale up is simple and predictable. Indeed, Eq. 1.1 is valid for any CCC column not involving the CCC column volume. It remains true whatever the total instrument volume, $V_M + V_S = V_C$, is. However the injection capacity, hence the productivity, both depend directly on V_C. The solvents used with small CCC columns are exactly the same as those used with a larger volume CCC column; hence Eq. 1.1 allows predicting accurately the sample peak positions. In contrast, scale up in solid/liquid chromatography is not straightforward since it is usually difficult to get the very same chromatographic particles that were used in the analytical columns to develop the large scale preparative column.

Figure 1.1 shows an example of school separation of benzyl-alcohol and p-cresol optimized on a small 5 mL CCC hydrodynamic column. The separation was scaled-up one thousand times using a 4.6 l column. In this work the authors have increased the flow rate, the injection volume and mass by three orders of magnitude as to keep the run time identical working with higher processed amount (g in the 4.6 L column) instead of milligram quantities injected in the 5 mL column (Sutherland et al. 2005).

1.3 Instrumentation

CCC provides a very simple yet productive way to purify complex mixtures working with solvent mixtures. Despite these assets, this technique is not widely used in industry because of the very simple reason: Eq. 1.1 shows that solutes are separated if and only if a large volume, V_S, of stationary phase can be retained in the CCC instrument. After several CCC design working with gravity such as the Craig machine or the so-called droplet CCC columns (Conway 2011; Berthod 2002), centrifugal forces were found to be the best way to retain a liquid stationary phase while the mobile phase is pushed through. Necessarily, the implementation of centrifugal forces comes with rotating parts, motors, belts or gears subject to failure, together with vibration, noise and security issues. These unattractive features had major consequences in the slow development of CCC in the industry. Early CCC instruments containing rotors were often poorly engineered and the large equipments available made method development both time- and sample-consuming.

In the past years, the CCC column design and technology made significant improvements. Reliable CCC columns allow for serious instrumental control and fast method development, with smaller columns and automated process control. The

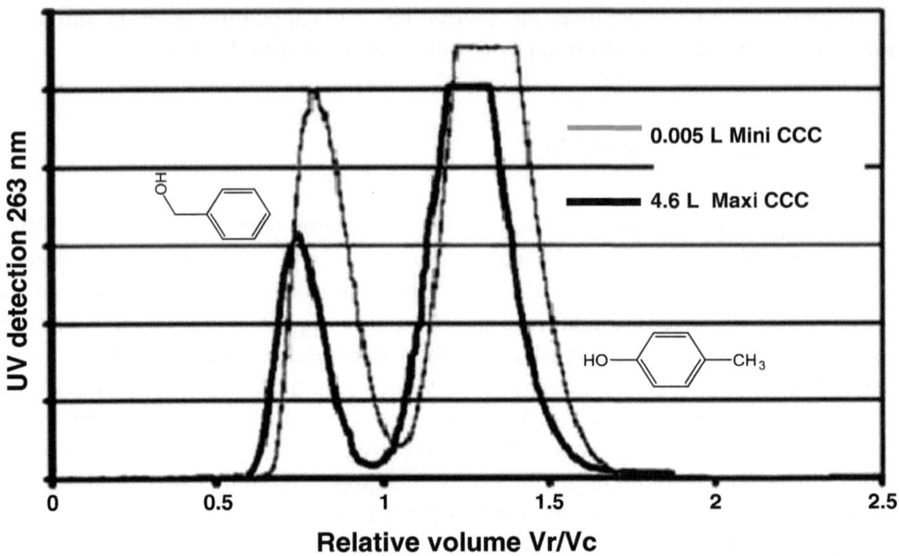

Fig. 1.1 Chromatograms resulting from the separation of a mixture of benzyl alcohol (*BA*) and p-cresol (*PC*) on a 5 mL-column (26 mg BA and 12 mg PC injected in 0.5 mL aqueous phase, F=1 mL/min lower aqueous mobile phase, 2100 rpm, Sf=64%) and a 4.6 L-column (24 g BA and 11.5 g PC injected in 460 mL of lower phase, F=6 L/min, 600 rpm, Sf=47%), both loadings are 10% of column volumes. Liquid system: heptane 4.7, ethyl acetate 0.3, methanol 1.7, water 3.3 (v/v) in the reversed phase mode with upper organic stationary phase and aqueous mobile phase in the descending or head-to-tail flowing direction. UV detection 263 nm. Adapted from (Sutherland et al. 2005) with permission of Taylor & Francis

latest commercially available instrumentations are presented in Table 1.1 along web sites and contact addresses. No doubt CCC can now integrate its full potential in industrial strategy for purification.

1.3.1 Hydrostatic and Hydrodynamic Instruments

From the numerous CCC prototypes developed during thirty years by Ito, two kinds of instruments were found economically viable and are described hereby (Conway 2011; Berthod 2002).

Sometimes called centrifugal partition chromatographs (CPC), the *hydrostatic* CCC instruments are build from stacked disks in which chambers are engraved in a radial direction, while being linked to each other by small ducts (Fig. 1.2 top). The disks are brought in rotation in a centrifuge. The disk rotation induces a constant radial centrifugal field (hydrostatic force). The liquid stationary phase (colorized in red in Fig. 1.2) is held in the chamber, while the mobile phase percolates through the channels. Chromatographic separation occurs in the chambers while the ducts remain filled with mobile phase. There are two possible working ways. In the *as-*

Table 1.1 Technical characteristics of commercially available CCC equipments

Model name	Column volume (mL)	Max rotation speed (rpm)	Max back-pressure (kg/cm^2)	Max flow rate (mL/min)	Typical handling size (g)
HYDROSTATIC CCC columns (Centrifugal Partition Chromatographs)					
Armen Instrument -Gilson—Z.I. de Kermelin, 16 rue Ampère, 56890 Saint Ave, France www.armen-instrument.com					
SCPC250	250	3000	100	50	5
TCPC250	250	3000	100	100	Continuous
SCPC1000	1000	3000	100	100	20
TCPC1000	1000	3000	100	100	Continuous
Kromaton-Rousselet-Robatel—BP 129 — 07104 Annonay, France www.kromaton.com					
FCPC analyt	50	2000	100	30	1
FCPC semi-prep	200	2000	100	100	5
FCPC prep	1000	2000	100	100	20
EverSeiko Corp. 4-39-5 Senzoku, Taitoku, Tokyo, Japan http://www.everseiko.co.jp					
CPC80	80	1500	60	10	1
CPC240	240	1500	60	20	5
CPC1400	1400	1100	60	80	20
HYDRODYNAMIC CCC columns					
Dynamic Extractions, 890 Plymouth road, Slough, Berkshire, SL14LP, UK http://www.dynamicextractions.com					
Mini	18	2000	~15	2	0.2
Spectrum[a]	140 (20)	1600	~15	10	2
Midi	940	1400	~15	100	20
Maxi	18000	800	~15	1500	1000
AECS-QuikPrep, 55 Gower street, London, WC1 6HQ, UK http://www.quattroprep.com					
IcMSPrep[a]	110 (30)	1200	~10	5	2
QuikPrep[a]	750	1200	~10	20	15
LabPrep[a]	2000	1200	~10	50	20
Tauto Biotech Ltd., 326 Aidisheng road, Zhangjiang Park, Shanghai, China http://www.tautobiotech.com					
TBE-20A	16	2000	~10	1	0.1
TBE-300B	260	1000	~10	4	2
TBE-1000A	1000	600	~10	10	10
CC-Biotech LLC, 9700 Great Seneca Highway, Rockville, MD 20850, USA http://www.ccbiotech.us					
STS-80	80	([b])	~15	([b])	2
STS-135	135	([b])	~15	([b])	4

[a] Device with several possible volumes (smaller volume in parentheses)

[b] STS stands for Spiral Tubing Support; the assembly must be placed in an existing hydrodynamic CCC rotor

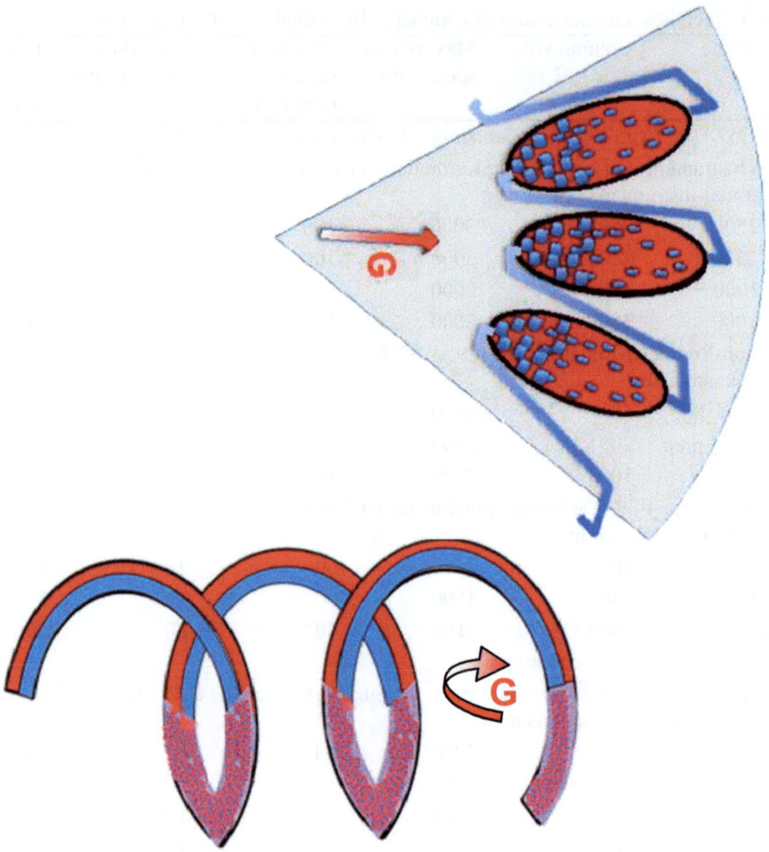

Fig. 1.2 Scheme of the two major CCC instruments. In red color: stationary phase, in blue color: mobile phase. Top—portion of a disk contained in a hydrostatic instrument showing the pattern of channel and ducts retaining the liquid stationary phase with a constant centrifugal field G; bottom—sketch of three turns of a coil contained in a hydrodynamic instrument. The variable centrifugal field G is high at the top inducing liquid phase decantation. It changes direction at the bottom inducing strong phase mingling

cending mode, the chambers are filled with the denser liquid stationary phase and the lighter mobile phase is pushed against the centrifugal field hence the ascending term by analogy with motion in the gravitational field is justified. In the *descending* mode, the stationary phase is the lightest and it fills the chambers, while the denser mobile phase is percolated through it in the direction of centrifugal forces. In both configurations, each geometrical channel is responsible for a significant back hydrostatic pressure proportional to the height of the liquid phase to cross and to the square of the rotating speed. In general, this design easily holds solvent systems, even with small density differences, but it suffers by the limited chromatographic efficiency due to the unused "dead" volumes introduced by the ducts containing only the mobile phase without any chromatographic exchanges.

Hydrodynamic CCC instruments are based on at least one coil mounted in a rotating frame. A gear arrangement generates a planetary rotation motion of the coil(s) with unsymmetrical and rotating centrifugal forces. Many complex designs have been attempted, mainly by the founder of the technique Yoichiro Ito and his group (Conway 1989). The most common instrument is called J-type hydrodynamic CCC. It is based on a planetary motion, with the helical coil rotating around its own axis, while the gear assembly rotates around a central axis, both axes being parallel and revolving at the same speed. During rotation, an oscillating or rotating force field is generated. When put in contact, the two immiscible liquid phases are decanting far from the central rotor axis with a strong centrifugal field (cumulating rotor and coil rotation), and mingling immediately after when passing close to the central axis where the centrifugal field due to coil rotation is subtracted to the rotor centrifugal field (Foucault and Chevolot 1998). The elevated number of mixing/decantation steps results in very good chromatographic efficiency, with high number of theoretical plates.

Due to the thread of the coil tubing, an Archimedeous pumping effect occurs, resulting in the suction of liquid phase at one coil end (called tail), while the pressure increases at the other end (called head). The denser phase of the biphasic liquid system used is always located at the tail side forcing the lighter upper phase to gather at the high pressure coil head. Consequently, if the mobile phase is the denser phase, it will be introduced in a *head-to-tail* direction in a similar way described as descending for the hydrostatic CCC column since it is against the Archimedeous pumping effect. If the mobile phase is the lighter phase, it will be pushed in opposite direction called *tail-to-head* or ascending. The main weakness of these J-type hydrodynamic CCC instruments may rely in liquid stationary phase retention. Some biphasic liquid systems, especially the polar solvent systems, are poorly retained due to high viscosity, low interfacial tension or small density differences between the two liquid phases. When a liquid system is well retained by hydrodynamic CCC columns, the observed chromatographic efficiency (peak sharpness) is higher than what can be obtained with the same liquid system and hydrostatic CCC columns of similar volume.

1.3.2 Liquid Systems

One has to keep in mind that the commercial hydrodynamic CCC instruments are in fact only empty tubes. As opposed to HPLC columns that are sold packed with chromatographic phases and are actually stationary phases, the mobile phase being selected independently, CCC columns have to be generated by the operator chemist prior to every analysis. The major advantage of this point is that a fresh column can be prepared on request, by simply bringing in contact two immiscible liquid phases. While HPLC deals with a limited number of stationary phases, CCC has to deal with the fine selection of the adequate biphasic liquid system to choose between unlimited numbers of possible systems. A CCC solvent system actually represents two chromatographic systems as either the lighter or the denser phase can be chosen as stationary phase. Any change in composition of either phase will affect the composition of the other phase. The choice of the solvent system fully depends on the sample and its polarity. Many reviews are dedicated to the selection of solvents (Foucault and Chevolot 1998; Hopmann et al. 2011; Yoichiro 2005; Berthod 1991). The range of polarity is wide, from least polar systems not containing water(e.g. heptane/methanol; heptane/acetonitrile, heptane/DMSO) to the most polar systems containing water in their two phases and called aqueous two-phase systems (ATPS), such as the phosphate/polyethyleneglycol (PEG)/water or dextran/PEG/water systems.

Standardization of solvent selection has been attempted, especially for natural product analyses. The heptane or hexane/ethyl acetate/methanol/water system is made by mixing selected volumes of these four solvents that forms two liquid phases. It was introduced by Oka in 1991 (Oka et al. 1991). These four solvents are appropriate having wide polarity difference, low viscosity, UV transparency, strong interfacial tension between the aqueous and organic phases and fast equilibration upon mixing associated with an acceptable impact on environment. Refining this system, a series of compositions with decreasing polarity was proposed. In the ARIZONA system, the alkane/ethyl acetate volume ratio is exactly the same as the methanol/water volume ratio. For example, the first composition, labelled A, is made with no heptane, 1 volume of ethyl acetate (ratio 0/1) and the same for methanol and water, that is, no methanol and 1 volume of water. The A composition is the binary ethyl acetate/water most polar system. The least polar composition Z is the binary waterless system heptane/methanol. This series of compositions (Table 1.2) was called the ARIZONA system because its compositions were referred to using the A to Z letters (Table 1.2). These compositions were found to cover the maximum range of polarity (Lu et al. 2009).

1.4 Counter Current Chromatography, a Green Process

1.4.1 Saving Solvents

In terms of ecological aspects, chromatographic techniques are characterized by a large consumption of organic solvents. Developing a greener process in chromatographic purification can consist in using smaller volumes of solvents. In this aspect, CCC appears at first sight as dealing with large volumes of solvents, but its high capacity results in an overall saving of solvents. A practical example will illustrate this saving. Spinetoram J is a powerful natural insecticide that needs to be purified from Spinetoram L, its fermentation by-product (DeAmicis et al. 2011). Using classical, solid phase based, liquid chromatography, a typical purification run by prep-LC of 2.5 g lasts 8 min with a flow rate of 800 mL/min, which leads to a consumption of 6.4 L of solvents per 2.5 g run. In a simple comparison, a CCC run lasts 140 min at a flow rate of 360 mL/min, i.e., *50 L per run but for a 111 g of crude load.* Since the

Table 1.2 Compositions forming the so-called Arizona liquid system

letter	Solvents as v/v/v/v				Upper phase composition % v/v				up/low	Density
	heptane	ethyl acetate	Methanol	water	Heptane	ethyl acetate	methanol	water	R	g/mL
A	0	1	0	1	0.0	97.0	0.0	3.0	0.88	0.903
B	1	19	1	19	5.3	94.7	0.7	2.6	0.92	0.920
C	1	9	1	9	10.3	85.8	1.4	2.5	0.965	0.878
D	1	6	1	6	13.7	82.0	1.9	2.4	0.96	0.870
F	1	5	1	5	17.2	78.1	2.4	2.3	0.95	0.862
G	1	4	1	4	19.7	75.6	2.6	2.1	0.95	0.856
H	1	3	1	3	25.8	69.4	2.8	2.0	0.945	0.842
J	2	5	2	5	30.8	64.4	3.1	1.7	0.91	0.831
K	1	2	1	2	36.2	58.8	3.5	1.5	0.88	0.818
L	2	3	2	3	46.1	49.9	3.0	0.9	0.84	0.795
M	5	6	5	6	54.6	41.8	2.8	0.7	0.80	0.777
N	1	1	1	1	62.5	34.4	2.6	0.5	0.70	0.760
P	6	5	6	5	69.4	28.2	2.0	0.4	0.69	0.746
Q	3	2	3	2	75.8	22.6	1.5	0.2	0.68	0.733
R	2	1	2	1	83.0	15.6	1.2	0.13	0.68	0.716
S	5	2	5	2	88.8	10.1	1.0	0.09	0.70	0.704
T	3	1	3	1	89.7	9.3	0.9	0.06	0.735	0.701
U	4	1	4	1	94.0	5.2	0.8	0.05	0.76	0.693
V	5	1	5	1	97.3	1.9	0.7	0.04	0.78	0.685
W	6	1	6	1	97.6	1.7	0.7	0.03	0.775	0.685
X	9	1	9	1	98.0	1.3	0.7	0.03	0.77	0.684
Y	19	1	19	1	98.0	1.2	0.8	0.02	0.71	0.684
Z	1	0	1	0	97.6	0.0	2.4	0.0	0.45	0.683

The upper over lower phase volume ratio, R, is related to the phase volume percentages as: lower phase percentage $= 100/(R+1)$; upper phase percentage $= 100R/(R+1)$; e.g. the phase ratio of A is $R = 0.88$ giving 53.2 % of lower phase and 46.8 % upper phase. Data at room temperature (22 °C ± 1 °C) taken from (Lu et al. 2009)

important point of view is productivity, the volumes of solvent per kg of purified compound must be compared for each technique. In this example, CCC uses *733 L* per 1 kg produced in 9 runs, while prep-LC uses *2560 L per kg produced in 400 runs*. Prep-LC needs three times more solvent volume than CCC.

The total time needed to perform the 9 CCC runs is at least 1260 min or 21 h; more likely 25 h including the setting time between runs. The 400 prep-LC runs make 3200 min or 53 h and more likely 60 h total time. Then the throughput of CCC is 40 g/h of purified Spinetoram compared to 17 g/h for HPLC. In short, for this particular example, CCC is able to produce a kilogram of purified Spinetoram 2.5 times faster than prep LC and using three-fold lesser solvent (DeAmicis et al. 2011).

This comparison cannot be generalized to any purification process, but it clearly demonstrates that in some cases, despite its negative image, CCC, based on liquid-liquid exchanges, can be a much greener process than conventional purification techniques based on liquid-solid phase exchanges such as prep LC.

Another interesting point in solvent management in CCC is the fact that the two liquid phases being non miscible, it is possible to recycle both mobile and stationary phases after each run. Recycling is compulsory when dealing with industrial purifications, due to economical reasons. In CCC, the majority of separations are achieved in isocratic mode. The collection of mobile phase in between chromatographic peaks affords direct recycling. The stationary phase itself can be extruded and monitored to check its purity before recycling.

Solvent systems are generally prepared by mixing the appropriate solvent proportions (Table 1.2) and allowing equilibration and the appearance of the two non miscible phases. But thanks to precise GC analyses of the composition of each phase, it is now possible to prepare each phase individually. Finally, the possibility to recover solvents in product fractions after evaporation and readjustment to final composition before reuse in CCC has been demonstrated (Garrard et al. 2007).

1.4.2 Improving Process Parameters

Another strategy to improve both economical and ecological aspects is by improving process parameters, in terms of mass factors (intensity, separation, efficiency and greenness) (Zhang et al 2011). The throughput can be improved by implementing multiple injections in the purification process, i.e. injecting while the previous separation is still running. One of the objectives to greatly improve throughput is to obtain online purification, meaning a continuous injection mode. A unique advantage of the CCC technique is that any one of the two liquid phases of the biphasic liquid system can be used as the stationary phase. It is even possible to switch the phase role during a separation. This is called working in "dual-mode" (Berthod 2002; Conway 1989). Using such phase role switch several times was called "multiple dual-mode" (Delannay et al. 2006). Multiple dual mode strategy consists in an elution based on frequent switching between mobile phase and stationary phase. Solutes that partition more in the upper phase will elute at one end while solutes that partition in the lower phase will elute at the other end of the column. Injections can be performed either in the middle of the CCC column (or technically speaking between two CCC instruments), as shown in Fig. 1.3 (Van den Heuvel et al. 2009) or at one end of the column as on Fig. 1.4 (Delannay et al. 2006). The injected solutes are moving back and forth, virtually increasing the chromatographic column length then allowing for complete separation.

1.4.3 Injecting Crude Samples

The liquid nature of the two phases allows the direct injection of crude or heavily loaded samples. Solubility can be achieved in either (mobile or stationary) phase or in a mix of both phases. Because no adsorption can occur and open tube clogging is limited, it is possible to load crude samples, limiting the sample preparation of natural products to a simple liquid extraction. For example, Armbuster et al. (Armbruster et al. 2001) achieved the purification of 64 crude plant extracts. Pushing this advantage to its limit, it was found possible to inject samples containing particulate matter (Sutherland 2007) as shown in Fig. 1.5.

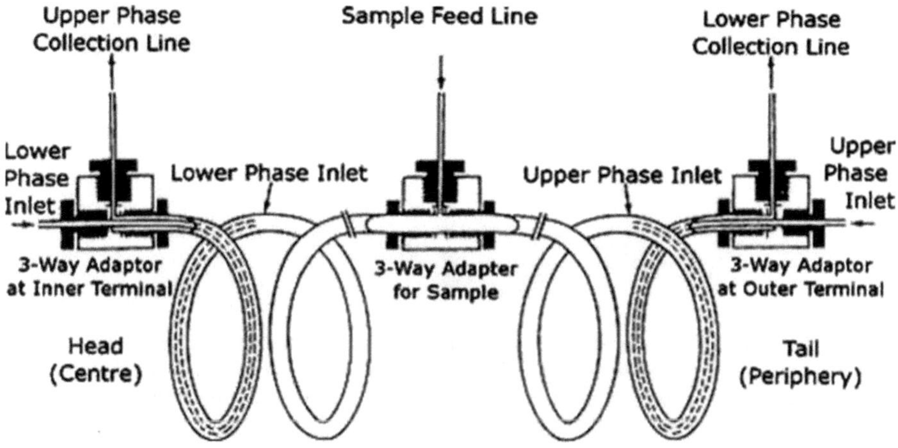

Fig. 1.3 Diagram showing the reduced bore tubing inserted one turn in both inlet and outlet of a hydrodynamic CCC coil to allow for collection of one liquid phase while entering the other phase. This design allows for a true countercurrent circulation of the two liquid phases. Reprinted from (Van den Heuvel et al. 2009) with permission of Elsevier.

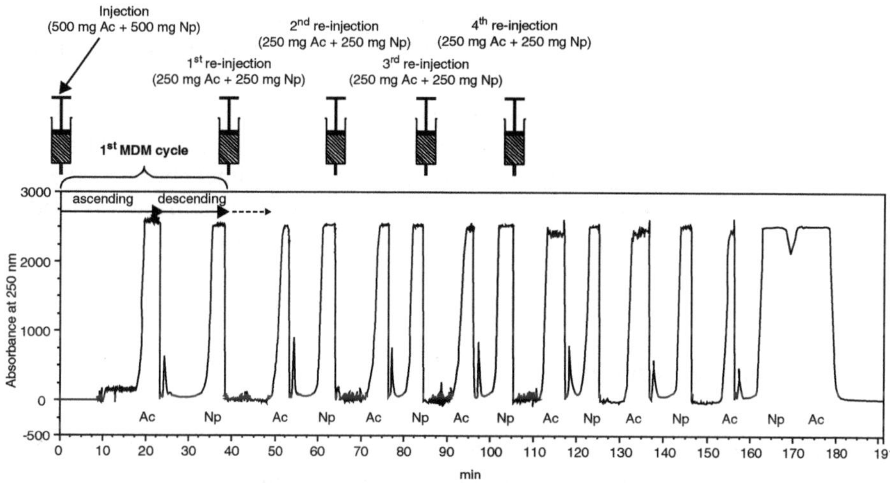

Fig. 1.4 Multiple dual-mode separation of acenaphthylene (Ac, 1.5 g) and naphthalene (Np, 1.5 g) using the heptane/acetonitrile liquid system. In ascending mode (or tail-to-head) heptane is the mobile phase and acetonitrile the stationary phase. 200 mL hydrostatic CCC column, rotor rotation: 1800 rpm; flow rate in both modes: 8 mL/min; UV detection 250 nm. Reprinted from ref (Delannay et al. 2006) with permission of Elsevier

Such concentrated injection is not possible in prep-LC since the bed of silica particles inside the LC column would act as a filter and be rapidly clogged by the particulate matter. This CCC development increases the greenness of CCC technique, as it can drastically reduce sample preparation before chromatographic purification.

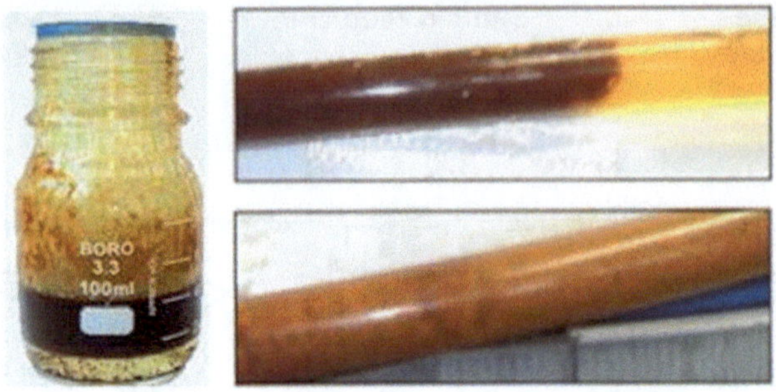

Fig. 1.5 Extremely concentrated solutions (*left*) can be injected in a CCC column. The right part of the figure shows a partial precipitation that occurred inside the injection loop during large scale natural product extract purification. The chromatogram developed normally. Adapted from (Freebairn et al. 2011) and Dynamic Extractions with permission

In another original example of use of the extraction power of CCC, Fedotov and Thiebaud (Fedotov and Thiébaut 2000) used the sludge sample in which thay had to extract components as the stationary phase in a hydrodynamic CCC column. The sewage sludge was introduced slowly in the column and set in rotation. The centrifugal force field was sufficient to maintain this highly loaded and polar medium with suspended particulates. The apolar mobile phase selected was a n-heptane-dichloromethane mixture that was slowly pumped through the sewage sludge placed inside the CCC hydrodynamic column. All polycyclic aromatic hydrocarbons (PAH) were extracted from the sludge and eluted according to their solubility in mobile phase. This example illustrates the potential of the CCC instrument besides standard chromatographic separation (Armbruster et al. 2001). Dichloromethane was used in this 2000 article. As will be exposed, nowadays the use of this solvent is strongly discouraged.

1.4.4 Greener Solvents

Water is the greenest solvent; it is also the most polar and universally available. It can be used for most CCC purifications involving some polar components. Its density makes the aqueous phase as the denser phase in most combinations with organic solvents producing a biphasic liquid systems (Berthod 2002; Conway 1989).

The chlorinated solvents have been very popular in CCC because these have excellent solvent properties and are denser than water. They provides lower organic mobile phase layer that could be run in the descending head-to-tail mode and are easier to handle with hydrodynamic column (positive pressure) than the ascending or tail-to-head mode (Archimedeous suction effect in tail-to-head mode). The methanol/chloroform/water system has been used for a long time and works efficiently with solutes of intermediate polarity.

According to the European Registration, Evaluation, Authorization and Restriction of Chemical (REACh regulation) that aims to improve the protection of human health and the environment, halogenated solvents have to be strictly avoided due to demonstrated environmental and toxicity concerns. Alternative solvents were found such as esters, ketones, nitriles, or ethers, to come close in polarity, but all these solvents are lighter than water producing upper organic phases.

New "greener" solvents are now being considered. For example, Faure et al. recently intriduced limonene, the orange essential oil as an alternative to alkane phase (Faure et al. 2013). However, the "greenness" of solvents in purification technique is an assessment that has to be carefully apprehended. One criterion for greenness is usually that the solvent is hardly evaporated, therefore limiting atmospheric pollution. For this purpose, ionic liquids are strongly recommended and have been studied with interest as new possible CCC solvent (Ruiz-Angel et al. 2007). However, the "greenness" of a solvent may be questioned when dealing with a purification technique, as the energy needed for its making and its removal from the purified CCC fractions are criteria that have to be taken into account in the global ecological impact of the process.

1.5 Counter Current Chromatography, a Tool for Green Chemistry Development

1.5.1 Natural Products

The high productivity of CCC offers an economical interest for the purification of natural products in order to produce active principles. Indeed, CCC has been used mostly in the study of Traditional Chinese Medicines (TCM). This technique makes it more profitable to extract active compounds from natural products such as plants than to synthesize the molecules from chemicals. This trend represents the major use of CCC in the industry (Marston and Hostettmann 2006), and especially in the TCM field (Sutherland and Fisher 2009).

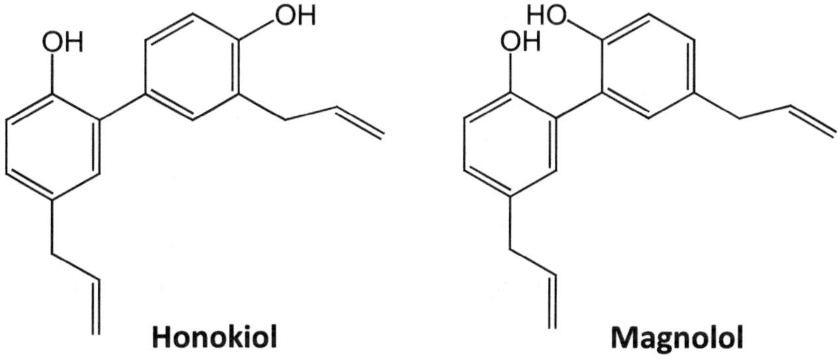

Scheme 1: the honokiol and magnolol isomer structures

As an illustrative example, the honokiol purification is presented. Honokiol is an anti-inflammatory active molecule found in the barks of Houpu, a TCM vegetable. An ethanol extraction of simply crushed Houpu barks results in the release of two isomers: honokiol and magnolol (Scheme 1). Because the separation of the two isomers was difficult to achieve in prep-LC, a CCC method using the Arizona **S** liquid system (Table 1.2) with 230 mL hydrodynamic instrument was developped. A throughput of 32 mg/h was obtained with run treating 150 mg of ethanolic bark extract in a two hour run producing 64 mg of 99.2 % pure honokiol using 100 mL of solvent (Wang et al. 2004). The honokiol value is so high that scaling up the published purification was profitable. With a large 4.6 L CCC hydrodynamic column, the Brunel UK team was able to use the same biphasic liquid system and fully optimize the purification (Chen et al. 2007). They showed that it was possible to process 43 g of bark extract with a mobile phase flow rate of 0.6 L/min producing 23.2 g of 98.6 % pure honokiol in about 20 min achieving a throughput of 51 g/h (1600 times higher, 51 g/h compared to 32 mg/h) with a CCC column only 20 bigger (4.6 L compared to 230 mL) and using seven times less solvent per produced gram compared to the initially proposed purification) (Chen et al. 2007).

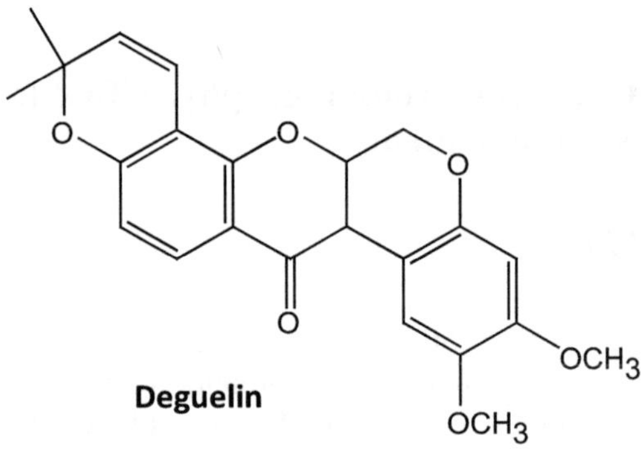

Scheme 2: Structure of Deguelin

Another example will show the purification of deguelin (Scheme 2), a natural flavonoid rothenoid present in a TCM plant. Deguelin is suggested to act as a potent anti cancer drug (colon and breast cancers). However, it is interesting, on both economical and time scale points of view, to evaluate natural deguelin in clinical trials before considering its industrial synthesis. Moreover, structural analogs present in the *Millettia pachycarpa* Benth plant containing deguelin have not been studied in terms of bioactivity. Separation and purification of deguelin and its analogues were studied and successfully performed by CCC using a 290 mL hydrodynamic column and the Arizona **Q** system (Table 1.2). It was possible to obtain 109 mg of 95 % pure deguelin, along with five analogues from 400 mg of plant extract. This result was obtained in 6 hours using 0.7 L of solvent but more than 100 mg of pure deguelin was enough for clinical preliminary studies and CCC did speed up the whole pharmaceutical process .

1.5.2 Solute Partition Coefficient Determination

The REACh regulation while force many changes in European chemical processes and likely sooner or later in the whole chemical worldUnsafe solvents will need to be replaced to conform to the more stringent regulation. In the quest of solvent substitutes, polarity is a key parameter. Solvent polarity is estimated using octanol/water partition coefficient, $K_{o/w}$, or other solvent distribution ratio between two liquid phases. $K_{o/w}$ coefficients were extensively used in quantitative structure-activity relationship (QSAR) studies (Hansch et al. 1995). Partition coefficient can be extrapolated or calculated using the compound molecular structure. However, when dealing with ionizable or dimerizable species, biases occur in computation so that a true measurement of the partition coefficient is more accurate than any calculation.

The classical shake-flask method for partition coefficient measurement with a two phase liquid system has a tedious procedure including the need of a very pure sample associated to accurate weighting, dissolution, centrifugation for hours and spectrophotometric or HPLC analysis of both phases. CCC provides a faster result in a simple operation (Berthod and Mekaoui 2011): the partition coefficient between the two phases is determined solely by the migration time monitoring of the sample peak using Eq. 1.1 to form:

$$K_{o/w} = (V_R - V_{aq}) / V_{oct} \tag{1.2}$$

using a stationary phase volume of octanol, V_{oct} saturated with 40 g/L water (2.2 M at 20 °C), and a mobile phase volume of aqueous phase, V_{aq} saturated in by 54 mg/L octanol (0.42 mM) (Berthod 2002). The $K_{o/w}$ partition or ratio is the most important distribution ratio, since it has been internationally recognized as an acceptable marker for hydrophobicity and solute ability for permeation through biomembranes. The CCC result is not dependent on the amount of stationary phase or on the volume of the column. While the sample may not be perfectly pure, it may even be possible to determine more than one solute partition coefficient at the same time considering the different peaks appearing.

For example, the octanol/water biphasic system that is easily retained in any CCC instrument, was used to determine three ketone $K_{o/w}$ coefficients in one single run (Ruiz-Ángel et al. 2011) as shown in Fig. 1.6.

Thus, CCC is most useful to determine the $K_{o/w}$ distribution ratio of ionisable compounds without any assumption (Berthod et al. 1999).

In CCC, it is critical to know that the experimental retention volume of a solute (its peak position) depends on the solute distribution ratio between the two liquid phases. The distribution ratio is defined as the total concentration of a solute in the stationary phase in all possible chemical forms divided by the total concentration of the solute in the mobile phase in all its possible chemical forms. This ratio is constant, and can be called a coefficient, if and only if there is no chemical change of the solute in both liquid phases. In many cases, chemical ionization, dimerization and other molecular association, the distribution ratio is not constant depending on

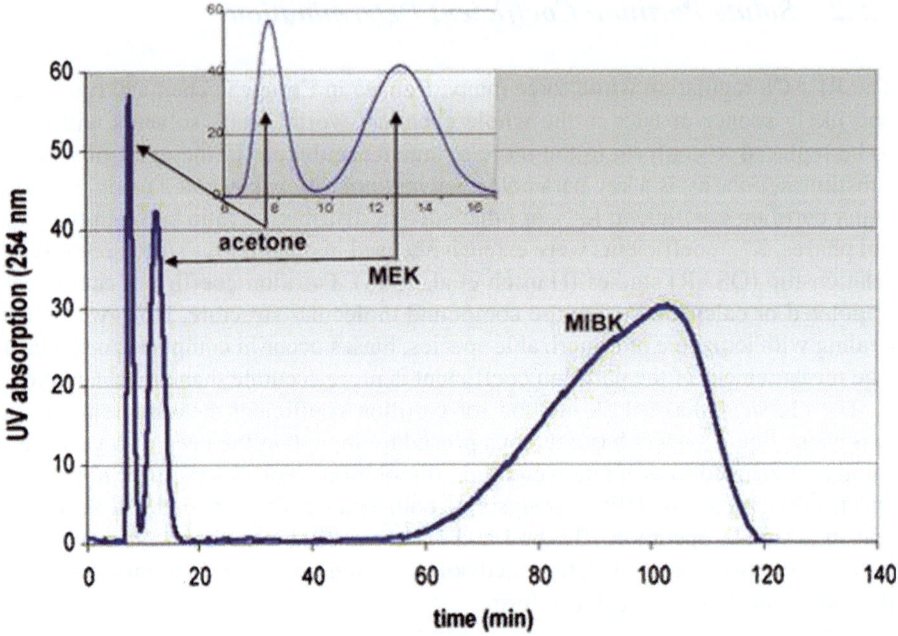

Fig. 1.6 Separation of three ketones for $K_{o/w}$ determination—Hydrodynamic CCC column of 75 mL, octanol stationary phase $V_{oct}=30$ mL, aqueous mobile phase at 8 mL/min in the descending (head-to-tail) direction $V_{aq}=45$ mL. The ketone peak retention volumes of 62, 101 and 800 mL allowed to compute the respective $K_{o/w}$ values of 0.56, 1.85 and 25.1 or log $K_{o/w}$ −0.25, 0.27 and 1.4. Reprinted with permission of Elsevier from (Ruiz-Ángel et al. 2011)

the change occurring between chemical forms. In those cases, it means that the CCC solute peak retention volume depends on the experimental conditions.

In this particular application, CCC is no more regarded as a purification tool, but more as an analytical tool. This is even increasing the interest of the technique, since suppliers start to downsize their instruments allowing for quicker $K_{o/w}$ determination, or, if detection is possible, for larger $K_{o/w}$ determination.

1.6 Conclusion

This review aims at describing the basis of Counter Current Chromatography (CCC) and showing its potential as a versatile and powerful purification technique working with two liquid phases and using only solvents.

It has been considered as a green process because of its lower solvent consumption and higher throughput capacity as compared to classical preparative liquid chromatography with powdered silica support. Moreover, this technique exhibits many attractive features (crude sample handling, no sample loss, versatility and high separation power) to be part of a greener chemical development based on natural product extraction.

The simplicity of separation mechanism and the development of more reliable instruments will undoubtedly ensure the spread of this technique in industry.

Acknowledgements AB and KF thanks the French National Center for Scientific Research (CNRS UMR5280—ISA) for continuous support.

References

Armbruster J, Borris R, Jimenez Q, Zamora N, Tamayo-Castillo G, Harris G (2001) Separation of crude plant extracts with high speed CCC for primary screening in drug discovery. J Liq Chromatogr Rel Technol 24:1827–1840

Berthod A (1991) Practical approach to high-speed counter-current chromatography. J Chromatogr A 550(0):677–693. doi:10.1016/s0021-9673(01)88574-5

Berthod A, Carda-Broch S, Garcia-Alvarez-Coque MC (1999) Hydrophobicity of Ionizable Compounds. A theoretical study and measurements of diuretic octanol–water partition coefficients by countercurrent chromatography. Anal Chem 71(4):879–888. doi:10.1021/ac9810563

Berthod A, Mekaoui N (2011) Distribution ratio, distribution constant and partition coefficient. Countercurrent chromatography retention of benzoic acid. J Chromatogr A 1218(36):6024–6030. doi:10.1016/j.chroma.2010.12.027

Berthod A (2002) Countercurrent chromatography, the support-free liquid stationary phase, Comprehensive Analytical Chemistry, vol 38. Elsevier, Amsterdam

Chen L, Zhang Q, Yang G, Fan L, Tang J, Garrard I, Ignatova S, Fisher D, Sutherland IA (2007) Rapid purification and scale-up of honokiol and magnolol using high-capacity high-speed counter-current chromatography. J Chromatogr A 1142(2):115–122. doi:10.1016/j.chroma.2006.09.098

Conway W (1989) Countercurrent chromatography–apparatus, theory and applications. VCH Publisher, Weinheim

Conway WD (2011) Counter-current chromatography: Simple process and confusing terminology. J Chromatogr A 1218(36):6015–6023. doi:10.1016/j.chroma.2011.03.056

DeAmicis C, Edwards NA, Giles MB, Harris GH, Hewitson P, Janaway L, Ignatova S (2011) Comparison of preparative reversed phase liquid chromatography and countercurrent chromatography for the kilogram scale purification of crude spinetoram insecticide. J Chromatogr A 1218(36):6122–6127. doi:10.1016/j.chroma.2011.06.073

Delannay E, Toribio A, Boudesocque L, Nuzillard J-M, Zèches-Hanrot M, Dardennes E, Le Dour G, Sapi J, Renault J-H (2006) Multiple dual-mode centrifugal partition chromatography, a semi-continuous development mode for routine laboratory-scale purifications. J Chromatogr A 1127(1–2):45–51. doi:10.1016/j.chroma.2006.05.069

Faure K, Bouju E, Suchet P, Berthod A (2013) Use of limonene in CCC: a green alkane substitute. Anal Chem 85:4644-4650. doi:10.1021/ac4002854

Fedotov P, Thiébaut D (2000) Direct extraction and separation of some polynuclear aromatic hydrocarbons (PAHs) from complex mixtures (Sewage sludges) by countercurrent chromatography J Liq Chromatogr Rel Technol 23(6):897–907. doi:10.1081/jlc-100101496

Foucault AP, Chevolot L (1998) Counter-current chromatography: instrumentation, solvent selection and some recent applications to natural product purification. J Chromatogr A 808 (1–2):3–22. doi:10.1016/s0021-9673(98)00121-6

Freebairn K, Johns D, Douillet N, Thickitt C, Vilminot E, Mathews B, Harris G, Hewitson P, Sutherland IA, Ignatova S (2011) Chromatography…but not as we know it. Chromatography Today (June):32

Garrard IJ, Janaway L, Fisher D (2007) Minimising solvent usage in high speed, high loading, and high resolution isocratic dynamic extraction. J Liq Chromatogr Rel Technol 30(2):151–163. doi:10.1080/10826070601077252

Hansch C, Leo A, Hoekman D (1995) Exploring QSAR—Hydrophobic, Electronic, and Steric constants. American Chemical Society, Washington DC

Hopmann E, Arlt W, Minceva M (2011) Solvent system selection in counter-current chromatography using conductor-like screening model for real solvents. J Chromatogr A 1218(2):242–250. doi:10.1016/j.chroma.2010.11.018

Lu Y, Berthod A, Hu R, Ma W, Pan Y (2009) Screening of complex natural extracts by countercurrent chromatography using a parallel protocol. Anal Chem 81(10):4048–4059. doi:10.1021/ac9002547

Marston A, Hostettmann K (2006) Developments in the application of counter-current chromatography to plant analysis. J Chromatogr A 1112(1–2):181–194. doi:10.1016/ j.chroma.2005.10.018

Oka F, Oka H, Ito Y (1991) Systematic search for suitable two-phase solvent systems for high-speed counter-current chromatography. J Chromatogr A 538(1):99–108. doi:10.1016/s0021-9673(01)91626-7

Ruiz-Angel MJ, Pino V, Carda-Broch S, Berthod A (2007) Solvent systems for countercurrent chromatography: An aqueous two phase liquid system based on a room temperature ionic liquid. J Chromatogr A 1151(1–2):65–73. doi:10.1016/j.chroma.2006. 11.072

Ruiz-Ángel MJ, Carda-Broch S, Levet A, Berthod A (2011) Hydrophobicity of ionisable compounds studied by countercurrent chromatography. J Chromatogr A 1218(36):6044–6052. doi:10.1016/j.chroma.2011.01.015

Sutherland I (2007) Recent progress on the industrial scale-up of counter-current chromatography. J Chromatogr A 1151(1–2):6–13. doi:10.1016/j.chroma.2007.01.143

Sutherland IA, Hawes D, Ignatova S, Janaway L, Wood P (2005) Review of progress toward the industrial scale-up of CCC. J Liq Chromatogr Rel Technol 28:1877–1891. doi:10.1081/jlc-200063521

Sutherland IA, Fisher D (2009) Role of counter-current chromatography in the modernisation of Chinese herbal medicines. J Chromatogr A 1216(4):740–753. doi:10.1016/ j.chroma.2008.11.095

Van den Heuvel R, Mathews B, Dubant S, Sutherland I (2009) Continuous counter-current extraction on an industrial sample using dual-flow counter-current chromatography. J Chromatogr A 1216(19):4147–4153. doi:10.1016/j.chroma. 2009.01.077

Wang X, Wang Y, Geng Y, Li F, Zheng C (2004) Isolation and purification of honokiol and magnolol from cortex Magnoliae officinalis by high-speed counter-current chromatography. J Chromatogr A 1036(2):171–175. doi:10.1016/j.chroma.2004. 02.073

Ye H, Chen L, Li Y, Peng A, Fu A, Song H, Tang M, Luo H, Luo Y, Xu Y, Shi J, Wei Y (2008) Preparative isolation and purification of three rotenoids and one isoflavone from the seeds of Millettia pachycarpa Benth by high-speed counter-current chromatography. J Chromatogr A 1178(1–2):101–107. doi:10.1016/j.chroma.2007.11.060

Yoichiro I (2005) Golden rules and pitfalls in selecting optimum conditions for high-speed counter-current chromatography. J Chromatogr A 1065(2):145–168. doi:10.1016/j.chroma.2004.12.044

Zhang M, Ignatova S, Hu P, Liang Q, Wang Y, Luo G, Wu Jun F, Sutherland I (2011) Development of a strategy and process parameters for a green process in counter-current chromatography: Purification of tanshinone IIA and cryptotanshinone from Salvia miltiorrhiza Bunge as a case study. J Chromatogr A 1218(36):6031–6037. doi:10.1016/j.chroma.2010.12.118

Chapter 2
Ion Size Exclusion Chromatohtaphy on Hypercrosslinked Polystyrene Sorbents as a Green Technology of Separating Mineral Elecyrolites

Maria Tsyurupa, Zinaida Blinnikova and Vadim Davankov

Abstract The review considers a new preparative method of separating concentrated solutions of mineral electrolytes into individual components by size exclusion chromatography on neutral nanoporous hypercrosslinked polystyrene sorbents NanoNets (Purolite International Ltd., UK and USA). Basic principles of the method as well as factors determining the selectivity of separations are discussed. Unprecedented effect of a spontaneous increase in the concentration of separated components is explained on the basis of the concept of ideal separation process. The unprompted partial resolution of inorganic salts into parent acids and bases is a logical consequence of the size exclusion mechanism of ion separation; at the same time, this resolution proves the correctness of our understanding of the separation mechanism. The review discusses briefly previous works in this field and true reasons for well-known "acid retardation" process, the process of separating mineral acids from their salts on anion exchange resins under conditions excluding ion exchange.

2.1 Introduction

Many enterprises face the challenge of utilizing waste aqueous concentrated mixed solutions of mineral acids, bases and salts. Such kind of by-products is formed on industrial manufacturing inorganic compounds, like mineral fertilizers or soda. Galvanic wastes of mechanical engineering and electronic industries as well as waste pickling solutions produced by metal treatment and hydrometallurgy also have to be processed. As a rule, these products represent acidic or basic solutions of various salts of iron, zinc, tin, nickel, cadmium, bismuth, titanium, copper, etc. The extraction of valuable metals from these rather concentrated acidic or basic solutions by means of conventional ion exchange is not promising for many reasons, including,

V. Davankov (✉) · M. Tsyurupa · Z. Blinnikova
Nesmeyanov-Institute of Organo-Element Compounds,
Russian Academy of Sciences, 119991 Moscow, Russia
e-mail: davank@ineos.ac.ru

Inamuddin, A. Mohammad (eds.), *Green Chromatographic Techniques*,
DOI 10.1007/978-94-007-7735-4_2, © Springer Science+Business Media Dordrecht 2014

but not limited to, an unsatisfactory selectivity of ion exchange in a number of cases, the shrinkage of resin beads in concentrated solutions, which diminishes the efficiency of column operation and the resin lifespan. Besides, a rapid consumption of resin exchange capacity on processing concentrated electrolyte solutions and, correspondingly, a frequently repeated regeneration of ion exchangers substantially increase the process cost. Moreover, every regeneration cycle produces mineralized waste solutions requiring additional care.

Several attempts have been made to resolve the above problem. A real success in the processing and/or separation of electrolytes by the "ion retardation" process was achieved with amphoteric "snake-cage polyelectrolytes" suggested by Hatch et al. (1957) in 1957 and defined as "cross-linked polymer system containing physically trapped linear polymer". Soon Dow Chemical Co. commercialized these materials under the trade name "Retardion" Dow Chemical Co. Retardion-550WQ2 (water content 41.6 %), a typical snake-cage amphoteric polyelectrolyte, was made by polymerizing vinylbenzyl-trimethyl ammonium chloride inside a sulfonated polystyrene-type cation exchanger Dowex-50 W×2 and had both strong basic and strong acidic functional groups. On a 100 mL column packed with this resin, a 20 mL sample (2.0 N in NH_4NO_3 and 1.6 N in HNO_3) was almost completely separated into its constituents (Hatch and Dillon 1963). Similarly, 1.98 N $FeCl_2$ was partially separated from 3.15 N HCl. In both cases, the acids resided in the column longer than the salts.

In 1958 in a theoretical study on the activity coefficients of electrolytes in the anion exchange resin phase (Dowex 1×10), Nelson and Kraus (1958) arrived at the conclusion that the separation of HCl from corresponding salts occurs, probably, due to a low activity of HCl in the phase of the anion exchanger. They reported that the breakthrough curve of 0.1 N HCl shifts to 1.5 bed volumes if the acid was dissolved in 5 N LiCl, while the acid retention rises to 8 bed volumes when the salt concentration increases to 16 N.

Five years later Hatch and Dilon (1963) described again the separation of excess acids from their salts on standard exchangers under conditions that exclude ion exchange. This separation contradicted the conception of "ionic exclusion", according to which all strong electrolytes are excluded from ion exchange resin phase due to Donnan's effect (Helfferich 1962; Wheaton and Bauman 1953). The strong basic anion exchange polystyrene-type resin Dowex-1×8 was found to function especially well, and the authors introduced a new term "acid retardation" defining the process in which an acid moves along the column slower than its salt. The acid retardation process can be performed at high flow rates and elevated temperatures. The process was optimized and since 1976 widely exploited by Eco-Tec Canada (Brown et al. 1997; Brown and Fletcher 1988), as well as by others (Götzelmann et al. 1987) on the industrial scale.

Concerning the mechanism of acid retardation, Hatch and Dillon (1963) note that various ideas or their combinations may be offered to explain the retention of strong acids (e.g. HCl) from their salt solutions on anion exchange resins (in the acidic, e.g. chloride form):

1. Electrolyte interactions, including "salting effects", the HCl being salted-out into the resin phase.
2. Interaction of protons of the strong acid with benzene rings of the resin matrix.

3. Association of strong acids to form ion pairs and non-ionized molecules in the resin phase, due to the low dielectric constant of that phase.
4. Possibly an entropy-increasing or energy-lowering effect of excess protons on the microstructure of the water inside the resin matrix.

However, having examined many experimental data, Hatch and Dillon (1963) concluded that "none of the ideas are proved by the data currently available".

During the 50 years that passed since the publication of these classical studies, several research groups examined acid retardation processes on ion exchangers, but new ideas have not been suggested for the explanation of the non-trivial phenomenon of electrolytes discrimination. Interestingly, a more or less adequate mathematical description of experimental data is possible both in terms of self-association of acid molecules in the homogeneous resin phase and/or association of acid molecules with thc functional groups of the resin (Soldatov et al. 2004) and in frames of a heterogeneous, two-phase model of the resin bead (Ferapontov et al. 1999). However, the first approach is based on the above ideas that have been formulated, considered and still rejected by the pioneers of the method. And suggesting heterogeneous structure for gel-type anion exchangers, supposedly composed of a swollen polymer and a separate phase of "free" water, lacks any experimental corroboration. In fact, several mathematical models may well fit experimental data into equations, but this fact does not prove the reality of the conceptions they are based on.

When analyzing the publications concerning acid retardation process, we arrived at the conclusion that the effective discrimination of an acid and its salt is conditioned on steric inaccessibility of narrow internal spaces in the bead of anion exchange resin to large hydrated metal cations, rather than any attractive interactions of the acid with styrene-divinylbenzene matrix or functional groups of standard anion exchangers. In other words, we believe that the separation of mineral electrolytes proceeds via mechanism of size exclusion chromatography.

Size exclusion chromatography separates particles, macromolecules and low molecular weight compounds in accordance with their size and ability to migrate into fine pores of column packing. Large particles that can diffuse only into the widest pores mostly reside in the interstitial space and hence are rapidly transported along the column by the flux of mobile phase. Smaller particles penetrate into stagnant zones of both large and small pores and reach later the column outlet. However, to exploit this method for separation of the simplest hydrated mineral ions (diameters of hydrated ions being around 6–10 Å), it is necessary to take advantage of a sorbent having small pores. The progress we made in separation of electrolytes was conditioned upon the choice of hypercrosslinked polystyrene as the column stationary phase.

2.2 Nanoporous Hypercrosslinked Polystyrene Sorbents

Davankov and Tsyurupa (Davankov et al. 1969; Davankov and Tsyurupa 1990) were the first to introduce the hypercrosslinked polystyrene sorbents as far back as 1969. 25 years later Purolite International Ltd. (UK, USA) started to produce these materials as Hypersol Macronet™ on an industrial scale. Nowadays, hypercrosslinked polystyrene sorbents are widely used in industrial adsorption processes (Tsyurupa 1995; Davankov et al. 2002; Davankov and Tsyurupa 2011) as well as in analytical chemistry. Thus, in environmental analysis they are used as the best sorbents for solid phase extraction (Purosep, Isolute, LiChrolut EN), because the hypercrosslinked sorbents efficiently take up and concentrate the trace amounts of both non-polar and polar organic compounds from aqueous solutions and air. It has also been reported (Proskurina et al. 2007) that distinct from the situation with other SPE-materials, many aromatic and polar organic substances can be concentrated on hypercrosslinked polystyrene even from aliphatic hydrocarbon solutions. Microbeaded hypercrosslinked polystyrene sorbents proved to be very promising column packing materials for high performance liquid chromatography. Since hypercrosslinked polystyrene is compatible with any kinds of organic and aqueous mobile phases, it is possible to conduct experiments in different HPLC modes using the same column (Sychov et al. 2004, 2009). It is interesting to note that non-functionalized hypercrosslinked polystyrenes exhibit affinity to some heavy metal ions such as mercury, bismuth, lead or silver which form labile complexes with the π-electron systems of the aromatic polymer (Tsyurupa et al. 2003).

The above theoretically interesting and practically useful properties of hypercrosslinked polystyrene are determined by the high accessibility of aromatic network fragments to small molecules. Indeed, hypercrosslinked polystyrene is obtained by intensive crosslinking polystyrene chains in the presence of a solvent that solvates well the polymer chains and prevents them from microphase separation (Davankov and Tsyurupa 2011). Reaction of a bifunctional crosslinking agent with polystyrene phenyl rings results in obtaining rigid open-network construction composed of mutually condensed and conformationally rigid meshes:

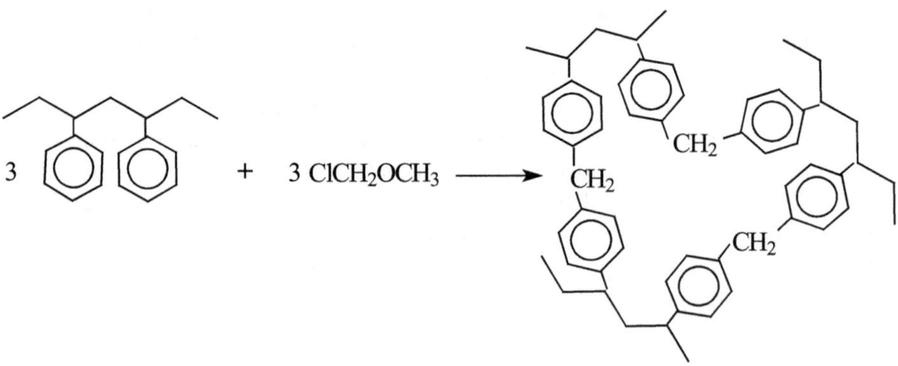

The material thus obtained represents a transparent single-phase polymer with a reduced density of about 0.5–0.7 g/cm^3 and high apparent specific surface area of 1000–2000 m^2/g. The unusually large free volume of hypercrosslinked network, up to 0.6 cm^3/g, as a matter of fact, is a true porosity of a new type, because the “pores” here correspond to small spaces confined by network meshes or, what is the same, to the openings between neighbor polystyrene chains, rather than to canals between dense polymeric walls. The maximum of the narrow "pore" size distribution corresponds to 1.5–3 nm depending on the conditions of network synthesis. The internal free space in the hypercrosslinked network is readily accessible to molecules of gases as well as liquids, irrespective of whether the latter strongly solvate the hypercrosslinked polystyrene network (typical solvents for linear polystyrene as toluene, dichloromethane) or display little affinity to polystyrene (precipitate polystyrene from its solutions, as hexane, methanol). Even water can occupy the network internal volume because water does not cause any shrinkage of the dry nanoporous material. Thus, the internal space in hypercrosslinked polystyrene network becomes easily accessible to molecules dissolved in water or organic liquids.

2.3 Brief Description of Chromatographic Experiments

The majority of experiments were performed using nanoporous hypercrosslinked polystyrene sorbents NanoNet (NN) manufactured by Purolite International. The sorbents were obtained by post-crosslinking swollen styrene-divinylbenzene copolymers containing chloromethyl groups via Friedel-Crafts reaction. These sorbents do not contain any unintendent functional groups capable of ion exchange. The content of residual chloromethyl and hydroxymethyl groups does not exceed 0.1 meq/g so that they do not play any noticeable role in separation of concentrated electrolytes.

Water pre-swollen beads of 0.3–0.5 mm in diameter were slurry packed in glass columns, mostly of 30 mL or 44 mL in volume (about 1 cm in diameter). In order to avoid tailing of chromatographic zones due to convective fluxes of a more dense (compared to water) solution of electrolytes, the latter was percolated through the column from its bottom upwards till the equilibration of the column with the feed solution completed. This experiment was called direct one. Afterwards, electrolytes were displaced from the column in opposite direction by the flow of pure water. This is reversed experiment. (In all figures the breakthrough curves resulting from direct experiments are separated by a gap from the elution part of the chromatograms). Under these conditions, the shape of fronts is determined only by the nonlinearity of phase distribution isotherms and longitudinal dispersion of concentration fronts. In the both direct and reversed experiments the effluent was collected in fractions of 1.2–1.5 mL. The flow rate was, on average, 1 mL/min. The concentration of electrolytes in each fraction of the effluent was then determined using appropriate volumetric analysis.

2.4 Dimensions of Hydrated Ions

To discuss the separation of electrolytes in frames of the ion size exclusion process, while accepting the idea that strong electrolytes completely dissociate into constituent ions, one needs to have information about the size of the individual ions in

Table 2.1 Hydration numbers of cations determined by different methods (Tanganov 2005)

Method of measurement	Hydration numbers of cations					
	Li^+	Na^+	K^+	Rb^+	Cs^+	NH_4^+
Electrophoretic mobility of ions	–	–	16	13	14	1
Self-diffusion coefficient	14	8.4	5.4	–	–	–
	–	–	22	–	–	–
	–	16.9	9.6	6.4	4.7	10.7
Rate of diffusion through membrane	5	3	4	–	–	–
Dependence of activity coefficient on concentration	7	3.5	1.9	1.2	–	–

Table 2.2 Hydration numbers of anions determined by different methods (Tanganov 2005)

Method of determination	Hydration numbers of anions			
	F^-	Cl^-	Br^-	I^-
Thermochemical method	–	4.4	4.8	5.3
Isothermal compression	4.5	2.0	1.8	1.5
Stoke's radius	5.5	3.9	3.4	2.8
NMR chemical shift	1.6	−1.1[a]	−3.2[a]	−7.5[a]
NMR relaxation time	9.9	13.2	16.2	21.8
IR-spectroscopy	6.0	6.0	6.3	9.1
X-ray diffraction	4.5	6	6	8.8
Neutron scattering	–	5.8	–	–

[a] Shift of the NMR signal to high field

aqueous media. There has been a great deal of debate for more than a century on the solvation of ions and the effect that dissolved ions have on the tetrahedral network structure of water. A central theme of that discussion has been the notion that different ions have "structure making" or "structure breaking" effects, depending on the charge and size of the ion. While potassium is regarded as being rather neutral in its effect on the structure of water (Hribar et al. 2002), the information collected so far for other mineral ions, even for the simplest ions, is very contradictive and inconclusive.

With respect to the number of water molecules comprising the hydration shell of an ion, static and dynamic methods of investigation show incompatible results. Static methods include thermochemistry (Samoilov 1957, 1972), NMR time relaxation (Fabricand and Goldenberg 1961), IR spectroscopy (Bergstrom et al. 1991), X-ray diffraction (Licheri et al. 1976; Biggin et al. 1984), or neutron scattering (Soper and Weckström 2006). These techniques provide information on the size of the hydration shell of ions in the bulk solution. Viscosimetrical determination of the Stokes radius (Marcus and Solvation 1985), measuring isothermal compression of solutions (Bocris and Saluya 1972), the rate of ion migration through a membrane as well as ion self-diffusion coefficients (Pinto and Graham 2004) are considered as dynamic methods, which characterize the moving ions. Tables 2.1 and 2.2 demonstrate how the results of the above measurements are contradictory. Unfortunately, we cannot judge which of them are more or less trustworthy. For that reason in our work we have used the list for sizes of hydrated ions from published

Table 2.3 Crystal radii (r_{cryst}) and effective radii of hydrated ions (r_H) (25 °C) (Nightingale 1959)

Ion	r_{cryst} Å	r_{hydr} Å	Ion	r_{cryst} Å	r_{hydr} Å
$(Me)_4N^+$	3.47	3.67	OH^-	1.76	(3.00)
$(Et)_4N^+$	4.00	4.00	F^-	1.36	3.52
$(n\text{-}Pr)_4N^+$	4.52	4.52	Cl^-	1.81	3.32
$(n\text{-}Bu)_4N^+$	4.94	4.94	Br^-	1.95	3.30
$(n\text{-}Pe)_4N^+$	5.29	5.29	I^-	2.16	3.31
H^+	(2.82)		NO_3^-	2.64	3.35
Li^+	0.60	3.82	ClO_3^-	2.88	3.41
Na^+	0.95	3.58	BrO_3^-	3.08	3.51
K^+	1.33	3.31	IO_3^-	3.30	3.74
Rb^+	1.48	3.29	ClO_4^-	2.92	3.38
Cs^+	1.69	3.29	IO_4^-	3.19	3.52
Ag^+	1.26	3.41	MnO_4^-	3.09	3.45
Tl^+	1.44	3.30	ReO_4^-	3.30	3.52
NH_4^+	1.48	3.31	CO_3^{2-}	2.66	3.94
Be^{++}	0.31	4.59	SO_4^{2-}	2.90	3.79
Mg^{++}	0.65	4.28	SeO_4^{2-}	3.05	3.84
Ca^{++}	0.99	4.12	MoO_4^{2-}	3.23	3.85
Sr^{++}	1.13	4.12	CrO_4^{2-}	3.00	3.75
Ba^{++}	1.35	4.04	WO_4^{2-}	3.35	3.93
Ra^{++}	1.52	3.98	$Fe(CN)_6^{4-}$	4.35	4.22
Mn^{++}	0.80	4.38			
Fe^{++}	0.75	4.28			
Co^{++}	0.72	4.23			
Ni^{++}	0.70	4.04			
Cu^{++}	0.72	4.19			
Zn^{++}	0.74	4.30			
Cd^{++}	0.97	4.26			
Pb^{++}	1.32	4.01			
Al^{+++}	0.50	4.75			
Cr^{+++}	0.64	4.61			
Fe^{+++}	0.60	4.51			
La^{+++}	1.15	4.52			
Ce^{+++}	1.1	4.52			
Tm^{+++}	0.9	4.65			
$Co(NH_3)_6^{+++}$	2.55	3.96			

work (Nightingale 1959). Since the radii of many hydrated cations and anions were estimated using measurements by the same method of electro conductivity, at least, their comparison is possible. Table 2.3 presents these values.

2.5 Separation of Electrolytes on Nanoporous Hypercrosslinked Sorbents

Owing to the high rigidity of the open-network structure of hypercrosslinked polystyrene, the latter is fully accessible to small analyte molecules not only in organic but also in aqueous media. Using this property of the hydrophobic nanoporous matrix, the behavior of inorganic electrolytes in a hypercrosslinked polystyrene matrix can

Fig. 2.1 Elution profiles for HCl and $CaCl_2$ (examined separately) from the chromatographic column containing NN-381. Experimental conditions: 44 mL column, 0.8 mL/min flow rate

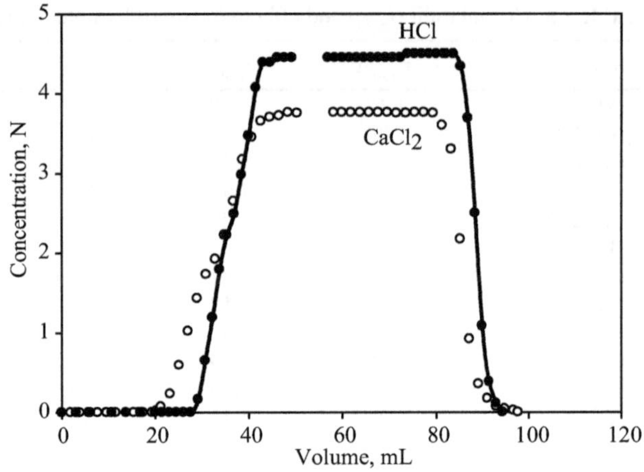

be studied. The column (44 mL) was filled with a suspension of 33.3 g water-swollen beads of sorbent NanoNet 381 in water. In this column, the interstitial volume amounts to 17.6 mL (40% of column volume, Manalo et al. 1959). Taking into consideration that one gram of the dry sorbent takes up 1.23 mL water, simple calculations show the total volume of water (mobile phase) in the column to be equal to 36 mL.

First of all, a 3.45 N solution of $CaCl_2$ was percolated through the column and then the salt was eluted with pure water. The same experiment was performed with 3.75 N solution of hydrochloric acid. Figure 2.1 demonstrates the trapezium-shaped combination of breakthrough and elution curves. In both the direct and reversed experiments the salt leaves the column sooner than the acid. Indeed, in the direct experiment the breakthrough of $CaCl_2$ occurs at 32.5 mL (measured at the center of concentration front) while the elution volume for HCl was 35.4 mL. The latter value practically equals the total volume of mobile phase in the column and so we have to arrive at the conclusion that, initially, the *neutral hypercrosslinked polystyrene sorbet NN-381 does not retain hydrochloric acid* and practically all pores in the sorbent are accessible to the acid, while $CaCl_2$ is partially excluded from the finest pores. In this experiment the fronts of $CaCl_2$ and HCl diverged by 7% of bed volume.

The situation changes dramatically when a binary concentrated solution of $CaCl_2$ in HCl is pumped through the same column (Fig. 2.2). Now, $CaCl_2$ emerges from the column much earlier than in the first experiment (22.7 mL against 32.5 mL) while HCl, on the contrary, retards and comes out with 43 mL. It looks like NN-381 retains hydrochloric acid only from its mixture with $CaCl_2$. In this chromatographic experiment with the binary solution the fronts of the two electrolytes are separated by one third of bed volume so that it is possible to get a fraction of pure $CaCl_2$ solution in the direct experiment and a fraction of HCl solution in the reversed experiment. The same good separations (sometimes even better) were obtained on percolating concentrated solutions of LiCl, KCl, $FeCl_2$ or $MgCl_2$ in concentrated hydrochloric acid. Moreover, in all cases an unusual fact engages our attention: the concentration of each separated component at column outlet exceeds their initial concentrations in the mixture.

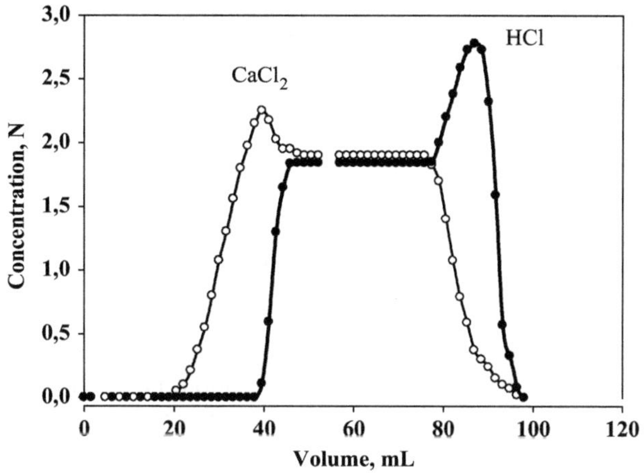

Fig. 2.2 Elution profiles for HCl and $CaCl_2$ (taken as a mixture) from the chromatographic column containing NN-381. Experimental conditions: 44 mL column, 0.8 mL/min flow rate

However, prior to explaining the difference in the above-described experiments we should answer the main question: why does the non-functionalized hypercrosslinked polystyrene separate ions?

Due to the principle of local electroneutrality, cations move along the column together with their anions. In the aforementioned examples (Figs. 2.1 and 2.2), the salt and the acid have the same chloride-anions. In that case the divergence of the salt and acid fronts may be conditioned upon one of two reasons, namely, (i) either the neutral polystyrene sorbent retains protons of the acid more strongly than it retains cations of alkali and alkali-earth metals, or (ii) different portions of sorbent internal volume are accessible to the acid and salt, i.e., they are excluded from the sorbent phase to different extents. Actually, the interaction of cations (including protons) with aromatic structures is well known (Ma and Dougherty 1997; Mecozzi et al. 1996). In hypercrosslinked polystyrene these interactions can be even enhanced because of the presence of condensed aromatic structures. For instance, anthracene-like fragments may readily emerge due to condensation of two chloromethylated styrene monomers during the synthesis procedure, occasionally followed by oxidative dehydration. Though, the amount of such proton sorption cites could not be high.

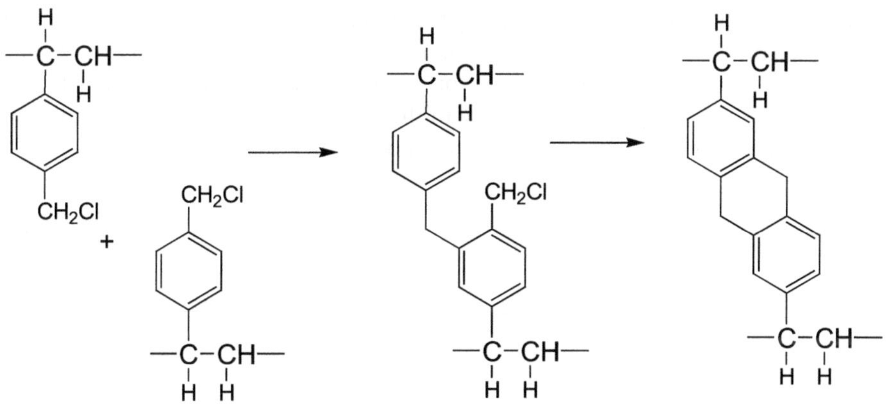

Table 2.4 Sorption of HCl, NaCl and NaOH on hypercrosslinked sorbent[a] NanoNet 381

Sorbate	Initial concentration, meq/mL	Equilibrium concentration,[b] meq/mL
HCl	1.364	1.360
HCl	1.417	1.425
NaCl	1.568	1.682
NaOH	2.050	2.108

[a] 50 g water-swollen sorbent, 100 mL solution, 3 days of periodical agitation at ambient temperature
[b] Calculated with a glance of water added with swollen polymer sample

On the other hand, partial exclusion of hydrated ions seems to be unavoidable. In hypercrosslinked sorbent NN-381 about half of pores have diameter of 1.5 nm so that one may expect the presence of smaller pores, comparable with the size of hydrated inorganic ions.

If hypercrosslinked polystyrene exhibits enhanced affinity to protons, it should be evident not only under chromatographic (dynamic) conditions as the appearance of the acid breakthrough front behind the front of its salt and the front of mobile phase, but also under static conditions as an increase in the acid concentration in the sorbent phase (when the sorbent stands in contact with a solution of acid). In the latter case, acid concentration in the outer aqueous solution must decrease. Table 2.4 shows the results of experiments conducted under static conditions. As can be seen, *the hypercrosslinked sorbent does not retain hydrochloric acid* since its concentration in the supernatant did not change after long contact of NanoNet 381 with the acid solution. Contrary to this, the concentration of NaCl and NaOH in similar experiments increased compared to their initial concentration. This implies that, as opposed to HCl and water molecules, certain part of total pore volume is inaccessible to NaCl and NaOH. In other words *cations of the salt and the base are partially excluded form sorbent phase.*

Cations of alkali and alkaline-earth metals are highly hydrated in aqueous solutions and their dimensions exceed that of hydrated chloride-anions (Table 2.3). Hydroxonium ion (H_3O^+) should also be hydrated in aqueous solutions, but it can easily appear in any place where a molecule of water is present under condition that the latter is connected to continuous aqueous phase by hydrogen bridges. Therefore, under chromatographic conditions proton does not move along the column because very fast shift of electrons along the chains of hydrogen bonds (see scheme below) can provide any required redistribution of charges in the aqueous phase, thus imitating the movement of protons and/or hydroxyls. This effect is responsible for extremely high electrophoretic mobility of these ions in aqueous solutions, as well as high electro conductivity of aqueous solutions of acids and bases.

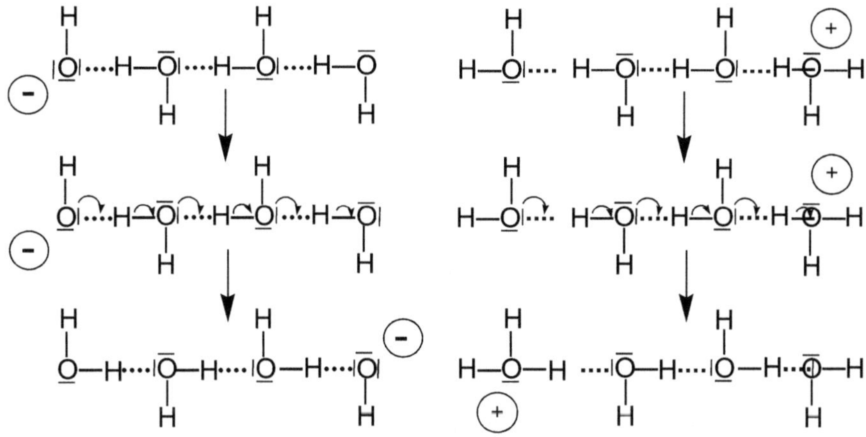

Thus, we can expect protons and hydroxyls to have zero eigen volume in aqueous solutions and experience no exclusion effects from any pores of the polymeric matrix. On the contrary, hydrated metal cations can reside only in sufficiently large pores while both large and smaller pores are accessible to chloride anions.

The above-mentioned distribution effects of metal and hydroxonium cations and chloride anions between the space of sorbent pores and interstitial solution clearly explain the effective separation of chloride salts from hydrochloric acid by exclusion chromatography. However, they do not explain the increase in the concentration of the salt and acid in the fractions in which these electrolytes appear separately from each other. The explanation of this phenomenon requires detailed consideration of specific features of exclusion chromatography.

2.6 Basic Features of Size Exclusion Chromatography

Exclusion chromatography is essentially employed in elution mode for the analytical-scale separation of large species. The application of this method to preparative separations requires closer consideration of its specific features which have usually escaped the attention of chromatographers (Davankov et al. 2005).

Species separated by size exclusion chromatography are transported along the column with the mobile phase, but move faster than the mobile phase. The totally or partially excluded analytes spend more time in the interstitial volume, i.e. in the moving part of the mobile phase, while all solvent molecules often diffuse into stagnant zones in all accessible pores of the packing. As a result, *the analytes quickly depart from their "mother sample solvent".* The excluded analytes constantly exploit the movement of those molecules of the mobile phase which happen to locate in the interstitial space, and they leave behind these solvent molecules at the moment the latter enter the porous space.

Let us consider the situation when a pump starts constantly delivering a solution of analytes into a column packed with porous material, which is the frontal mode of chromatography. Large analytes arrive at the column outlet and emerge in the effluent, leaving behind both smaller particles and mother sample solvent. At the moment the front of this sample solvent arrives at the outlet, the column becomes equilibrated with the feed mixture. At that moment, certain portions of excluded particles have already left the column with the effluent. Therefore, *a size exclusion column, being equilibrated with the mixture under separation, always incorporates liquid with reduced concentrations of the excluded species.* Even at a prolonged pumping, concentration of these species within the column will never reach to that of the feed. This fully corresponds to the name of the process: larger species are *excluded* from the packing, i.e., from the column. The larger the molecules of the analyte, the lower is their concentration in the sorbent-immobilized portion of the mobile phase and, hence, within the column as whole.

Though in frontal mode of chromatography all analytes emerge in the effluent in the sequence of their decreasing sizes, the analytes' concentrations in the effluent are usually expected to equal their concentration in the feed. Indeed, irrespective of the linear velocity of species along the column, the number of species emerging in one volume of the effluent must be equal to that simultaneously introduced into the column with same volume of the feed. This statement is also valid for the case of elution of all species from the "analyte-saturated", that is feed-equilibrated column. Most important is that *frontal size exclusion chromatography does not cause dilution of the solutes*. Of course, we assume here that all fronts are sharp and there is no noticeable dispersion at the borders of rectangular (or slightly trapezoidal) chromatographic zones.

Another special feature of size exclusion chromatography is that the process of *interaction between the solutes and the stationary phase is non-stoichiometric.* For this reason, the fundamental concept on exchange capacity or adsorption capacity of the column is no more applicable, and the concentration of the sample solution can be selected arbitrary. Only the solubility of sample components in the mobile phase can set a threshold for their concentrations. Naturally there is a limit for the volume of the sample that can be separated in one chromatographic cycle; theoretically, this is the total volume of pores in the column packing. This separation would correspond to a complete exclusion of larger species from all pores while smaller species freely enter even smallest pores of the packing. (It will be shown later that in the practice much larger sample volumes can be successfully applied).

It should be noted here, that all the above-emphasized features of size exclusion chromatography principally distinguish this technique from all common variants of adsorption, distribution or ion exchange chromatography. In the latter techniques, the analytes move through the column *slower* than the mobile phase (since they are partially retained by the stationary phase); the total concentration of the analytes in the column equilibrated with the feed solution is *higher* than the concentration of the initial feed (since analytes additionally accumulate on the stationary phase); the analytes eluting with pure mobile phase from the column appear in the effluent at concentrations *lower* than in the feed (since removing the retained analytes from the "analyte-saturated" sorbent requires additional amounts of the mobile phase); and the columns are easily *overloaded* (since the exchange capacity of every sorbent is a limited value).

2.7 Conception of "Ideal Separation Process"

In order to explain the unprecedented fact of a spontaneous increase in the concentration of electrolytes separated *via* ion size exclusion chromatography (Fig. 2.2), we would like to point out that preparative chromatography is only one of many approaches to separation and that all the approaches should be considered from viewpoint of involving additional auxiliary components in the separation process. Let us use the term "*ideal separation process*" to define any separation that *does not add* any supplementary matter to the separated fractions of the initial mixture (Tsyurupa and Davankov 2004; Davankov 2005). An automatic consequence of isolating (i.e., removing) one component from the initial mixture in such an ideal separation process would be the inevitable concentration increase of all other components in the remaining reduced volume of the mixture. When applied to distillation, crystallization, sieving and other similar separation methods, this statement is so self-evident, that it never needs to be formulated. On the other hand in chromatography, transportation through the column packed with an adsorbent usually dilutes the initial sample with large amounts of the mobile phase, so that the very idea of observing a *self-concentration effect for all analytes because of their separation* seems unrealistic at the first sight.

Indeed, one has always dealt with the retention of analytes on stationary phases. This retention decreases the analyte concentration in the moving zone of the mobile phase and requires additional amounts of the mobile phase to elute the retained portion of the analyte from the stationary phase. Known cases of peak compression in chromatography are mostly coupled with the displacement of the adsorbed analyte (or analytes) by an auxiliary component of the mobile phase (a displacer or a mobile phase modifier). In order to act like this, the latter must be adsorbed on the stationary phase even stronger than the displaced analytes. Therefore, in frontal chromatography process one may obtain weaker retained analytes in enlarged concentrations, however, peak compression can never be reached for a stronger retained component, in particular, for the displacer (Guioshon 2005).

The exclusion chromatography basically differs from all other modes of chromatographic techniques in the sense that the analytes are *not retained* by the column packing and therefore, do not need any special displacer or additional portions of the mobile phase in order to be eluted from the column. In accordance with the size exclusion mechanism the dilution of separated fractions is no more unavoidable. (The dilution can be minimized to the diffusion effects at the front and tail of the analyte zone). *The absence of any supplementary matter in the frontal size exclusion chromatography process relates it to the above defined ideal separation process.*

In this case the effect of self-concentration of salt and acid observed in experiments (Fig. 2.2) also becomes self-evident. Hydrated ions of calcium chloride which, in reversed experiment, move faster than acid ions depart the zone of mixed solution thus liberating a corresponding space in this zone. In exclusion chromatography an additional volume of water does not fill thus liberated space. Instead of this, the zone of hydrochloric acid, that is located behind the zone of calcium chloride, gets narrower thus leading to a definite increase in the concentration of the acid. Similarly, in direct experiment, $CaCl_2$ which passes through large pores with higher velocity than HCl moving slower through both large and small pores, gets rid of the acid and concentrates, as well.

As follows from the conception of ideal separation process, the larger molar volume of a component that is removed from the mixture and the higher was its concentration, the larger space is liberated in the initial mixture and the stronger is the effect of self-concentration of the remaining component (Guioshon 2005).

2.8 Selectivity of Electrolyte Separation Process

Figure 2.3 depicts chromatogram obtained on percolating the solution of 3.8 N $CaCl_2$ in 3.7 N HCl through the activated carbon D 4609 obtained by pyrolysis of hypercrosslinked polystyrene. The chromatogram is typical; elution fronts of the salt and the acid in the both direct and reversed experiments are widely separated. The effect of self-concentration of two components is clearly expressed.

As a quantitative characteristic of the electrolyte separation process, we suggest to use the breakthrough volumes measured at the middle of concentration wave height, $BV_{0.5}$. In direct experiment the difference Δ_1 between the fronts of two separated electrolytes divided by bed volume has to characterize the selectivity of separation of a faster moving component. The selectivity of separating the slower moving component Δ_2 is calculated by the same way in the reversed experiment as the difference between two $BV_{0.5}$ (also expressed in bed volumes). Note, these values are not necessary equal. In the example given in Fig. 2.3 the selectivity of separating calcium chloride is $\Delta_1=0.27$ while the selectivity of separating hydrochloric acid is $\Delta_2=0.28$; at that, the increase in concentration, $\lambda=C_{max}/C_o$, of separated hydrochloric acid amounts to 1.50 while $\lambda_1=1.35$ for the salt.

Since we understand now that the separation of electrolytes occurs via size exclusion mechanism, we can answer another important question: what factors determine the selectivity of separation? We should keep in mind that a quick shift of electrons along the chain of hydrogen bonds between water molecules can generate a positive or negative charge in any place of the aqueous phase where the corresponding charge is required for the maintenance of local electroneutrality. For this reason the velocity of HCl migration along the column is determined by the velocity of movement of chloride-anion rather than that of proton. Proton easily emerges in the vicinity of the moving anion and, probably, does not restrain the movement of the anion. In the same fashion, the movement of a base MeOH is entirely determined by the velocity of movement of the cation M^+, while hydroxyl-anion just has to be present near to the cation.

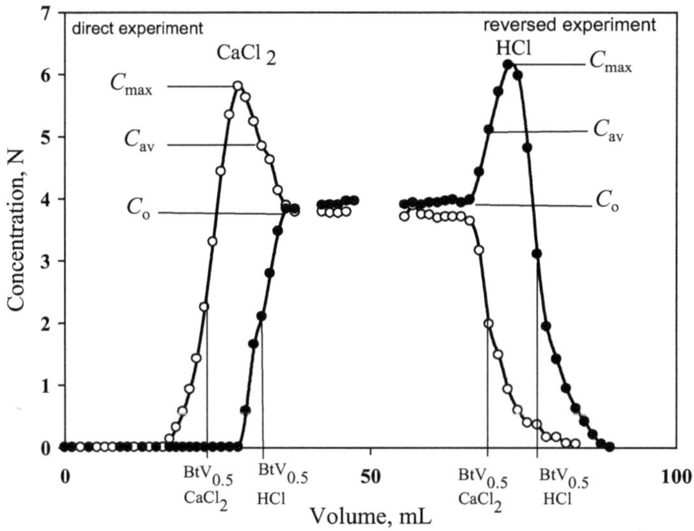

Fig. 2.3 Separation of 3.8 N calcium chloride and 3.7 N hydrochloric acid on D4609 carbon. Column volume: 30 ml; flow rate: 0.6 mL/min. Feed solution (50 mL) delivered from the bottom upward, then, the eluent (water) delivered from the top of the column downward

As regards the movement rate of a metal chloride, it is essentially determined by *the rate of movement of the largest constituent ion* (Davankov et al. 2005) which is excluded from a significant portion of the sorbent pore space and therefore is forced to migrate faster compared to its smaller counter ion, making the latter follow itself. The radius of hydrated Ca^{2+}ion is 4.12 Å (Table 2.2), larger than that of Cl^-, 3.32 Å. For this reason, anion Cl^- migrates immediately behind its cation without slowing down the overall rate of salt movement.

It logically follows from the above that the selectivity of separating two electrolytes MeCl and HCl is determined by the difference in size of hydrated Me^+ and Cl^- ions, i.e. the largest ions of the two electrolytes, rather than the difference between Me^+ and H^+ions, which actually conditions the distinction between two electrolytes. This pretty non-trivial explanation is confirmed by the separation of pairs NaCl-HCl and NaCl-NaOH. For the first pair (Fig. 2.4) the selectivity of separating fronts is determined by the difference in size of hydrated ions Na^+ (r_H=3.85 Å) and Cl^- (r_H=3.32 Å). This difference is not large but still it is sufficient to get good separation, Δ_1=0.27. In the second pair NaCl-NaOH (Fig. 2.5) sodium ion is the largest for the both electrolytes and so their separation in the direct and reversed experiments is insignificant. If we would compare the sizes of Na^+ and H^+ in the first pair of electrolytes (NaCl-HCl) and Cl^- and OH^- in the second pair (NaCl-NaOH), we could not explain the above great difference in the separation results. As opposed to the pair NaCl-NaOH, the separation of fronts for Na_2SO_4-NaOH amounts to 0.17 in the direct experiment (Fig. 2.6). In this case the difference in dimensions of

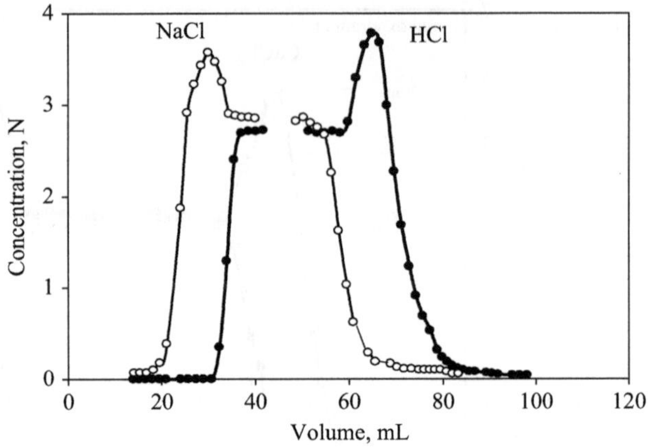

Fig. 2.4 Separation of 2.8 N sodium chloride and 2.7 N hydrochloric acid on D4609 carbon. Experimental conditions: 28 mL column, 1 mL/min flow rate

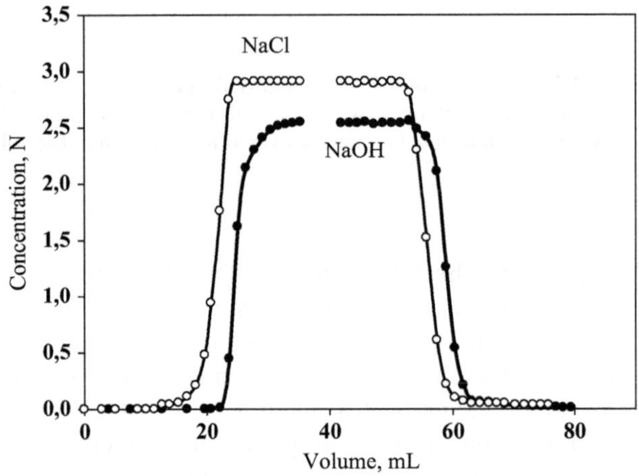

Fig. 2.5 Separation of 2.9 N sodium chloride and 2.5 N sodium hydroxide on D4609 carbon. Experimental conditions: 28 mL column, 0.8 mL/min flow rate

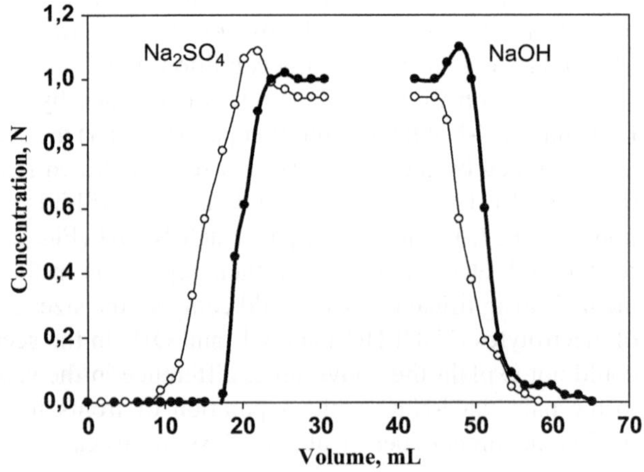

Fig. 2.6 Separation of 1 N Na_2SO_4 and 0.8 N NaOH on NN-381. Experimental conditions: column 28 mL, flow rate 0.8 mL/min

sulfate-anions (3.79 Å) and sodium cations (3.58 Å) determines the better selectivity of separation.

2.9 Influence of the Electrolyte Concentration on the Selectivity of Separation and Extent of Self-Concentration

In order to understand the surprising dependence of selectivity of the chromatographic process and self-concentration of separated electrolytes on their concentration in feed solution, first of all, we studied the behavior of model electrolytes $CaCl_2$ and HCl, taken separately and together, on contacting their solutions with water-swollen hypercrosslinked polystyrene sorbent NN-3881 under static conditions (Davankov et al. 2009). For this purpose 20 mL solution of known concentration was added to 5 grams of water-saturated sorbent beads. After the equilibrium was established, the distribution of electrolytes between the water phase that fills sorbent pore space and bulk aqueous solution was determined by titration of the latter. The phase distribution coefficient k was calculated as the ratio of electrolyte concentrations in these two parts of aqueous solution. Contrary to ion exchange resins, hypercrosslinked polystyrene sorbents do not change their volume in concentrated solutions of salts and acids of any concentration, which substantially facilitates calculation of phase distribution coefficients of electrolytes. Figures 2.7 and 2.8 depict the results obtained.

As can be seen in Fig. 2.7, at concentrations ranging from 0.5 to 4 N the phase distribution coefficient of HCl is close to unity. It means that this electrolyte neither adsorbs on the polymer, nor experiences any steric exclusion effects from the pore space and that HCl may reside in any pore which is accessible to water molecules. Interestingly, when the concentration is less than 0.5 N, the acid tends to concentrate in bulk solution (k_{HCl} decreases to 0.915 at the solution concentration of 0.09 N). Contrary to this, at very high equilibrium concentrations k_{HCl} exceeds unity reaching to the value of 1.13. The latter fact indicates that at high concentrations hydrochloric acid tends to accumulate in the sorbent pores. The dependence of k_{CaCl2} on

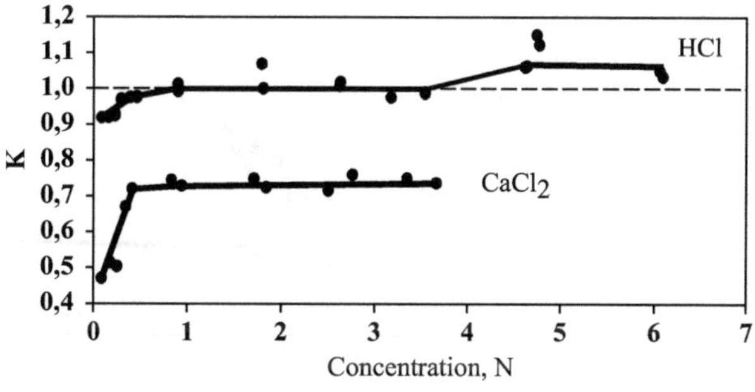

Fig. 2.7 Phase distribution coefficients for HCl and $CaCl_2$ (determined separately) between the pore volume of NN-381 and the bulk solution as a function of the concentration of the latter at equilibrium

the equilibrium concentration of the salt in bulk solution has the same form, although at the concentration of 0.09 N only a half of sorbent pores remain accessible to strongly hydrated calcium chloride. In the range of concentrations from 0.5 N to 4 N, the salt occupies three fourth of pore space ($k_{CaCl2}=0.75$). In more concentrated solutions, the phase distribution coefficient becomes equal to unity. It is difficult to explain the unusual character of changing phase distribution coefficients by any other reason than the decrease in extent of ion hydration with rising concentration of electrolytes in aqueous solutions. Unfortunately, in literature there is no trustworthy information concerning the influence of ion concentration on its hydrated size, however, it is legitimately to assume that ions of $CaCl_2$ and HCl are hydrated to the largest extent in the most diluted solutions. For this reason a certain portion of the smallest pores in hypercrosslinked sorbent proves to be inaccessible in dilute solutions even to chloride ions of the acid. The still greater increase in dimensions of hydrated calcium ions in highly diluted solutions also changes the behavior of calcium chloride; at low concentrations only half of internal space of stationary phase remains achievable to the salt. In concentrated solutions, on the contrary, the ions of $CaCl_2$ and HCl experience shortage of hydrating water molecules and effective size of ions decreases. Therefore, all pores of stationary phase become accessible to partially dehydrated calcium chloride.

It is quite interesting that phase distribution coefficients change radically when the two electrolytes simultaneously appear in a mixed solution (Fig. 2.8). Both the electrolytes tend to be excluded only in very diluted solutions. With equilibrium concentration increasing to about 4 N, the phase distribution coefficient of the acid rises from 1.0 to 1.57, while that of the salt diminishes from 0.75 to 0.45. To understand the above pattern, we should take into account the fact that only those chloride-anions that belong to HCl can reside in narrow pores, whereas in wide pores and interstitial space chloride-anions are present that belong to both HCl and $CaCl_2$. It inevitably leads to the emergence of a sharp gradient of chloride-anions (and total electrolyte) concentrations between small and large pores. This concentration gradient forces additional Cl^- ions to migrate into small pores (with simultaneous appearance of equivalent protons therein). As a result, HCl accumulates in small pores and partially in larger ones which were initially also accessible to $CaCl_2$ ions. The salt, on the contrary, is displaced to greater and greater extent into larger

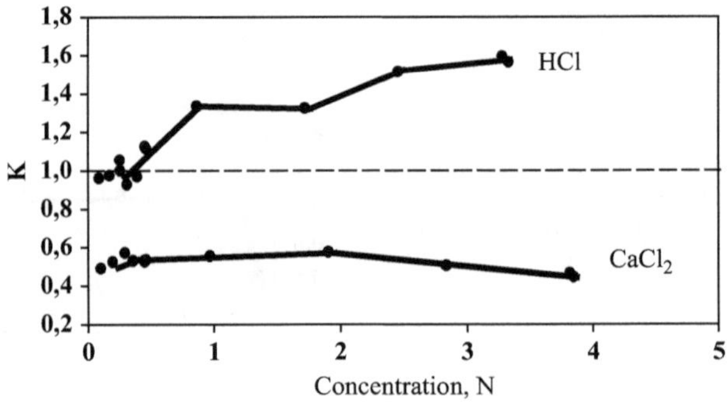

Fig. 2.8 Phase distribution coefficients for HCl and $CaCl_2$ (taken as a mixture) between the pore volume of NN-381 and the bulk solution as a function of their concentration in the bulk solution at equilibrium

pores and interstitial volume, causing there an increase in salt concentration. Thus the concentration of chloride ions (and, consequently, a total ionic strength of solutions) levels out in all parts of the aqueous phase. In other words, the tendency of system to level out the concentration of common chloride ions in all parts of aqueous solution results in the effect of acid self-concentration in the polymeric phase and self-concentration of the salt in bulk solution. This remarkable phenomenon permits to attain a noticeable separation of electrolytes even under static conditions.

In all probability, the decrease in the hydration numbers of calcium ions and, correspondingly, the decrease in their effective dimensions with the rising of $CaCl_2$ concentration, and *vise versa,* the growth of ion dimensions when concentration of HCl solution drops, result in a diffusive character of the leading breakthrough front of $CaCl_2$ and a noticeable compression of the trailing part of HCl elution zone (Figs. 2.1–2.4). On pushing the mixture of $CaCl_2$ and HCl through the column, the front of acid in the direct experiment is sharper than that of the salt. Here, the acid appears at the column outlet (later than calcium chloride) against the background of high $CaCl_2$ concentration. Under these conditions the hydration degree of HCl ions, and, hence, their size, becomes small and, presumably, invariable; breakthrough front of HCl is only distorted by normal diffusion effects. On the contrary, in the reversed experiment pure water follows the zone of the acid. Effective size of retarding HCl ions rapidly grows with dilution, and they are forced to move faster along the column. As a result, the boundary between the zone of acid and pure water gets narrower. As to the tail of $CaCl_2$ zone in the reversed experiment, the salt comes out of the column against the background of raising acid concentration. Therefore, the effective size of the last $CaCl_2$ ions decreases, their movement along the column slows down and so the edge front of calcium chloride becomes tailed.

Thus, the tendency of any system to level out the concentration gradients, in particular that of the ion common for a given pair of electrolytes, substantially increases selectivity of separation processes. Without any exception, this factor functions especially productive when smaller ions are presented in the mixture at a lower concentration compared to that of larger ions. The minor component is forced to concentrate and stay within stagnant thin pores until the major partner is removed from the column with water wash.

However, one has to bear in mind that hydration shells of ions are labile constructions and easily change with ionic strength, i.e. concentration of solutions. Informative examples present mixtures of calcium and potassium chlorides. In crystals, ionic radius of K^+ cation is noticeably larger than that of Ca^{2+}. However, in diluted aqueous solutions (0.17 N), calcium is stronger hydrated, as a smaller species with a greater electric charge, and it elutes from a NN-381 column ahead of potassium. Only 36 % of pore volumes of that sorbent appear available to Ca^{2+} in dilute solutions, compared to 65 % accessible to K^+. In concentrated mixtures, the situation reverses because calcium ion suffers intensive dehydration while potassium ion possesses less water molecules to loose. For this reason, it turns impossible to achieve a self-concentrating effect in an attempt to resolve a mixture concentrated in $CaCl_2$ (5 N) and diluted in KCl (0.5 N) so that both salts leave the column with the same volume. Indeed, at concentrations as high as 5.5 N, 89 % of

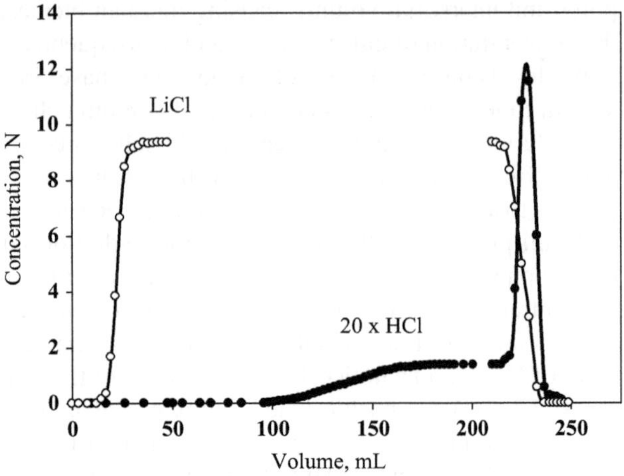

Fig. 2.9 Elution profiles for a mixture of 9.3 N LiCl and 0.07 N HCl from the chromatographic column with NN-381. The concentration of HCl on the plot is presented after multiplication by a factor of 20

pore volume becomes available for Ca^{2+} while this value for K^+ increases to 84 % only. For similar reasons, unsatisfactory selectivity was also achieved on separating such mixtures as 1.5 N $Ca(NO_3)_2$ - 1,5 N KCl (Δ_1=0.08), 1.5 N $MgCl_2$ - 1.5 N KCl (Δ_1=0), 4 N $CaCl_2$ - 0.5 N NH_4Cl (Δ_1=0.06). In these metal ion combinations selectivity of chloride salts discriminations does not exceed 0.1 in diluted solutions while it drops to almost zero with increasing concentration.

Such ion dehydration phenomena are barely examined at present, which makes attempts for predicting the separation efficiency by the size exclusion chromatography based on published ion diameter data less reliable.

2.10 "Acid Retardation", "Base Retardation" and "Salt Retardation" Phenomena

The tendency of systems to level out concentrations of the ion common to a given pair of electrolytes in all parts of the mobile phase, not only assists the self-concentration of electrolytes in different parts of the aqueous phase, but also increases substantially the selectivity of chromatographic separation process (Davankov et al. 2005; Tsyurupa et al. 2006). Figure 2.9 represents the chromatogram obtained when a mixture of concentrated (9.3 N) LiCl and diluted (0.07 N) HCl percolates through the porous material. A strong gradient of common chloride ion concentration "salts out" small ions H^+ and Cl^- of the acid into smallest pores where it accumulates until its concentration there becomes comparable to that of the major electrolyte LiCl in the interstitial liquid. Therefore the breakthrough of HCl in direct experiment occurs very late, after nearly 4 bed volumes of feed solution. Compared to the initial mixture, the acid concentration in the effluent rises herewith by a factor of 8.5 on eluting with pure water in reversed experiment. In this case the acid breakthrough volume exceeds by far the column volume. Certainly, such a process may be called "acid retardation" despite the fact that this effect has nothing in common with any attractive interactions between the acid and column packing.

On the contrary, if we separate a solution containing small amounts of LiCl and high concentration of the acid (Fig. 2.10), the salt fails to force its major partner, HCl into small pores, so that no self-concentration effect operates for HCl and the selectivity of separation merely reflects the difference in the size of Li^+ and Cl^-.

Even more interesting are the results of separation of LiCl from LiOH. In a diluted to about 0.5 N mixture, all constituent ions can be expected to exist in fully solvated state. The largest hydrated ion is lithium, common to the both electrolytes, and it must largely determine the migration rates of both LiCl and LiOH fronts. In accordance with the size exclusion principle, indeed, the two fronts arrive at the column outlet before the hold-up volume and with a minimal separation of the above pair salt-base, only $\Delta_1=0.1$ bed volumes (Fig. 2.11). This selectivity of salt-base differentiation rises by a factor of about twenty when LiCl is taken in the high concentration of 3.5 N (Fig. 2.12). Now, the selectivity amounts to more than two bed volumes ($\Delta_1>2$) and LiOH behaves as a strongly retarded component. Again, the accumulated base elutes with pure water in the form of a sharp concentrated peak (Tsyurupa et al. 2008).

The above discussed "acid retardation" and "base retardation" in the "immobilized liquid phase" could be related to the so-called "salting-out" effects. However, this term is hardly applicable to the case of "salt retardation", the first example of which was demonstrated by a successful removal of small amounts of NH_4Cl from a concentrated brine of $(NH_4)_2SO_4$ (Fig. 2.13). This practically important problem arises in the caprolactam manufacturing where large amounts of sulfuric acid are converted into ammonium sulfate used for the preparation of a crystalline fertilizer. The new process of ion size exclusion chromatography on nanoporous NN-381 resin allowed an effective purification of very large volumes of concentrated sulfate

Fig. 2.10 Chromatogram of a mixture of 0.25 N LiCl and 4.4 N HCl on a column with NN-381. Experimental conditions: 28 mL column, 1 mL/min flow rate. The concentration of LiCl on the plot is presented after multiplication by a factor of 10

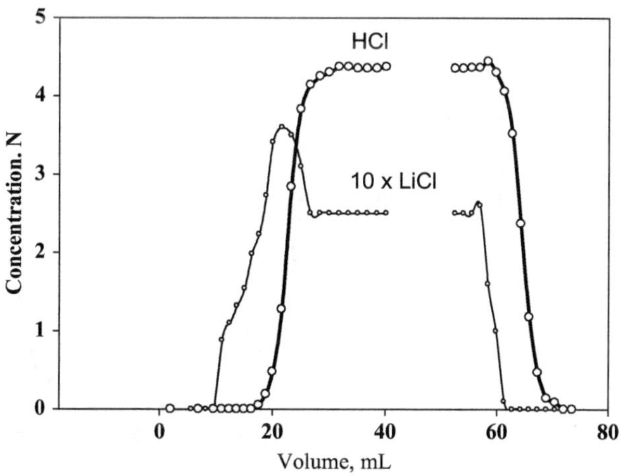

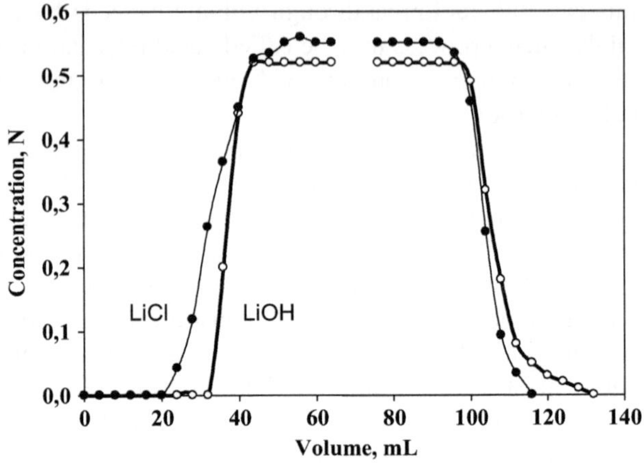

Fig. 2.11 Separation of 0.5 N LiCl from 0.53 N LiOH on NN-381. Experimental conditions: 30 mL column, 1 mL/min flow rate

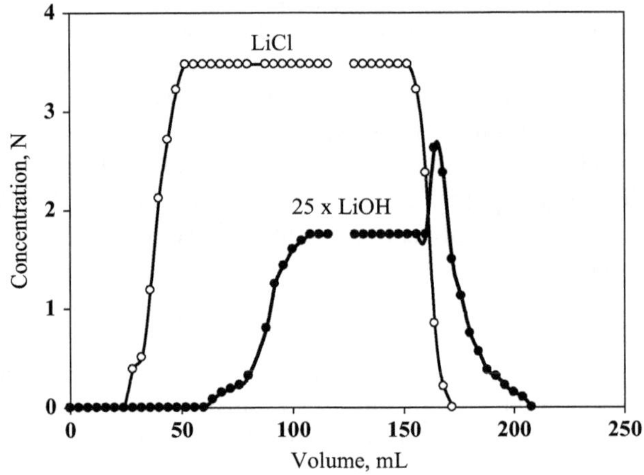

Fig. 2.12 Separation of 3.5 N LiCl and 0.07 N LiOH. Experimental conditions: 30 mL column, 1 mL/min flow rate. The concentration of LiOH on the plot was multiplied by a factor of 25

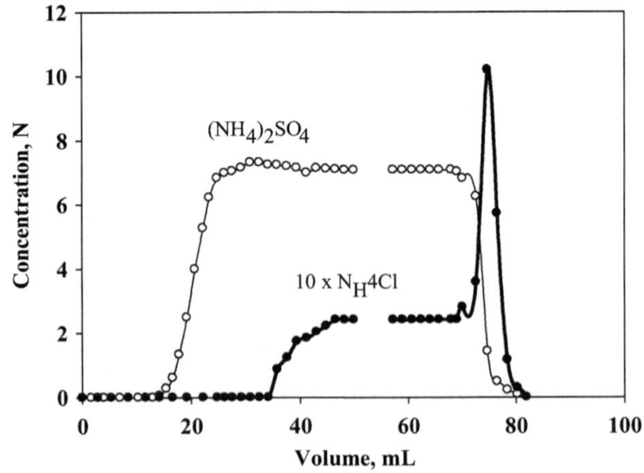

Fig. 2.13 Separation of 1 % NH_4Cl and 40 % $(NH_4)_2SO_4$. Experimental conditions: 28 mL column, 1 mL/min flow rate. The concentration of ammonium chloride is shown after multiplication by a factor of 10

brine, due to the fact that small ions of NH_4^+ and Cl^- are efficiently squeezed into and "retained" in finest pores of the sorbent (Tsyurupa et al. 2008). We consider this "salt retardation" process as a convincing proof of our interpretation of the mechanisms of the new electrolyte separation process.

2.11 Other Convincing Proofs of Separating Electrolytes via Exclusion Mechanism

Although we have every reason to state that the discrimination of electrolytes proceeds via ion size exclusion process, the next three examples finally leave no doubts in the correctness of our perception of the separation mechanism.

In the case of the first experiment, 28 mL column was filled with the beads of *dry* activated carbon D 4609 previously prepared by pyrolysis of a hypercrosslinked polystyrene. Opposite to polystyrene NN-381 resin, dry carbon can be directly wetted with water. The aqueous 0.5 N solution of $(NH_4)_2SO_4$ was thus pumped trough the dry stationary packing. After equilibration with the salt solution during the direct experiment, the electrolyte was displaced from the column with water-immiscible organic solvent, namely, *n*-butanol/chloroform mixture. Sulfate anions are large, $r_H = 3.79$ Å and cannot diffuse into the majority of small carbon pores which are accessible only to water molecules. Tending to move along the column faster than water molecules but at the same time having no possibility to depart from the mobile phase, ammonium sulfate accumulates on the water-air boundary, resulting in self-concentration in the first fraction of the effluent by a factor of 2.3 (Fig. 2.14). For the same reason, the salt concentration in the last aqueous fractions of the effluent advancing the organic eluent front, drops so that the very last aqueous fraction contains no salt, at all. Only size exclusion mechanism can explain such redistribution of the salt within the plug of water phase that passed through the column.

Second experiment involves the exchange of ions in electrolyte mixtures (Blinnikova et al. 2009). Figure 2.15 shows the results of chromatography of a mixture consisting of four different ions, obtained by dissolution of $MgCl_2$ and K_2SO_4. The mixture was actually separated into fractions composed of $MgSO_4$ and KCl by sending it through a column with water-swollen beads of NN-381. Figure 2.15 demonstrates the plots obtained by selective quantitative analysis of all four ions. The two largest ions of the mixture, Mg^{2+} and SO_4^{2-}, migrate through the column ahead of smaller ions, $K^+$and Cl^-. Since the two initial salts were taken in equivalent proportions, the elution fronts of the pair of larger ions coincide; the same is true for the pair of smaller ions. In this way, one pair of salts clearly converts into a new pair of salts.

In another experiment (Fig. 2.16), small ions of the above mixture were taken in deficiency compared to the two larger ions by preparing a mixed solution of 4.6 N in $MgSO_4$ and 0.1 N in KCl. As can be expected, a substantial increase in the total concentration (4.7 N) of the electrolytes results in markedly improved selectivity

Fig. 2.14 Redistribution of ammonium sulfate in aqueous solution passed through the column with carbon D 4609 and then eluted with *n*-butanol/chloroform (1:1 by volume) mixture

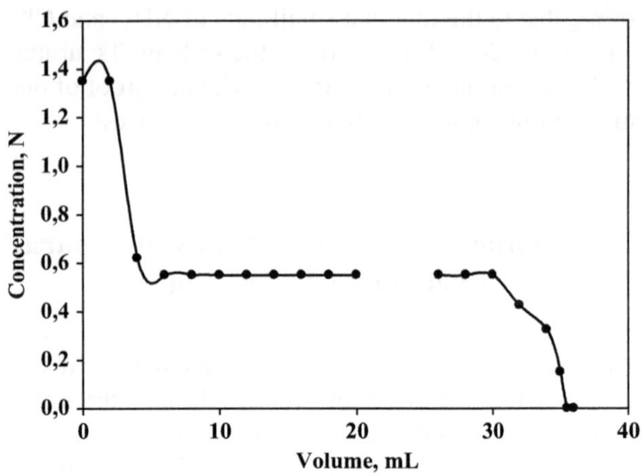

Fig. 2.15 Elution profiles for $MgCl_2/K_2SO_4$ mixture from chromatographic column with MN-381. Experimental conditions: 44 mL column, 0.8 mL/min flow rate

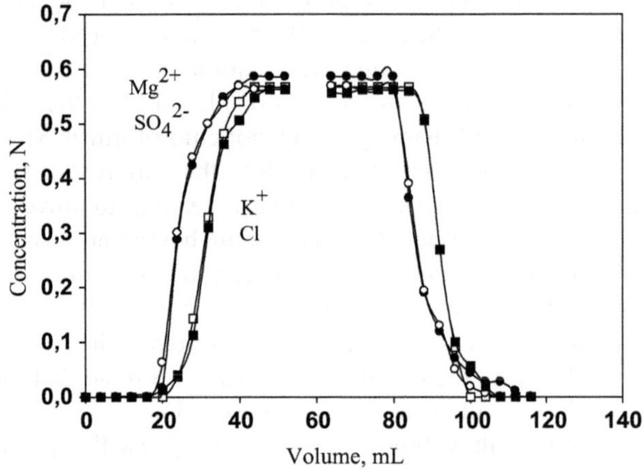

Fig. 2.16 Elution profiles for a 4.6 N $MgSO_4$/0.1 N KCl mixed solution from the column containing NN-381. Experimental conditions: 36 mL column, 0.8 mL/min flow rate

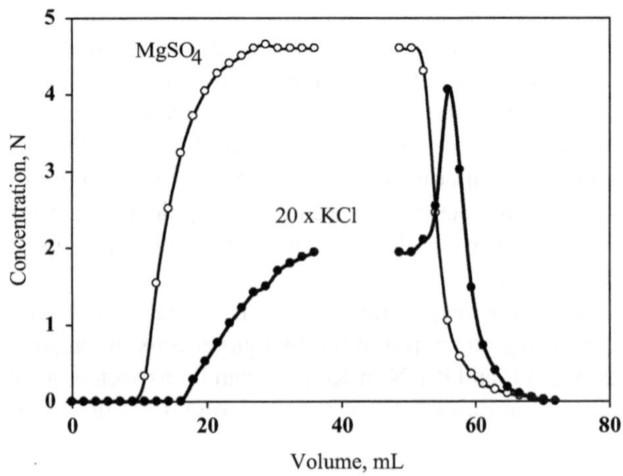

of separation, due to the operation of the concentration gradient factor. Importantly, the difference in the osmotic pressures between the water compartments in smaller and larger pores is now generated by entirely different ions (Mg^{2+} and SO_4^{2-} in large pores versus K^+ and Cl^- in small pores), rather than by a concentration gradient of a common ion (Cl^-), as was the case in all previously considered systems. This example implies that the general thermodynamic origin of all above discussed self-concentration effects which accompany acid, base or salt "retardation" phenomena, consists of the size differentiation-caused difference in the osmotic pressures or, which is actually more correct, the difference in the water activity in different parts of one single aqueous mobile phase.

Finally, the third vivid example illustrating the action of exclusion mechanism of electrolyte differentiation, is the separation of salts into parent acids and bases. If the sizes of the cation and anion of a given salt differ, they must tend to move through the nanoporous column packing with different velocities. The incipient separation of the cations of the salt from their anions is facilitated by a simultaneous emerging of protons and hydroxyls near corresponding ions, which would prevent formation of charged micro domains or charged zones in the moving aqueous phase. In other words, the salt in a process of size exclusion chromatography can be expected to gradually resolve into the corresponding acid and base (Davankov et al. 2006).

In order to test this prediction, 3-mL portions of different aqueous salt solutions were sent through conventional chromatographic columns packed with nanoporous hypercrosslinked polystyrene or poly(divinylbenzene). The column effluent was conducted through a flow-cell of a pH-meter. As can be seen form the initial drop of the effluent's pH, followed by its increase over the neutrality (Fig. 2.17), this tendency to resolve is indeed resulting in a partial separation of the zones of anions and cations, facilitated by the cooperation of protons and hydroxyls. The latter are always present in an aqueous phase, although in small concentration of 10^7 M (at pH 7 and room temperature), allowing the anions and cations to depart from each other, without the formation of electrically charged zones.

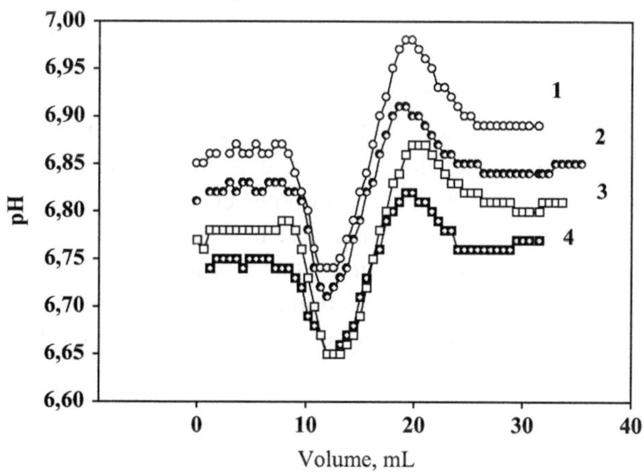

Fig. 2.17 Resolution of K_2SO_4 on a poly-DVB column. Column, 28 mL; probe: 3 mL of (1) 0.015 M, (2) 0.025 M, (3) 0.035 M, (4) 0.05 M solutions; flow rate: 1.2 mL/min of the 0.015 M KCl background electrolyte. After [181]

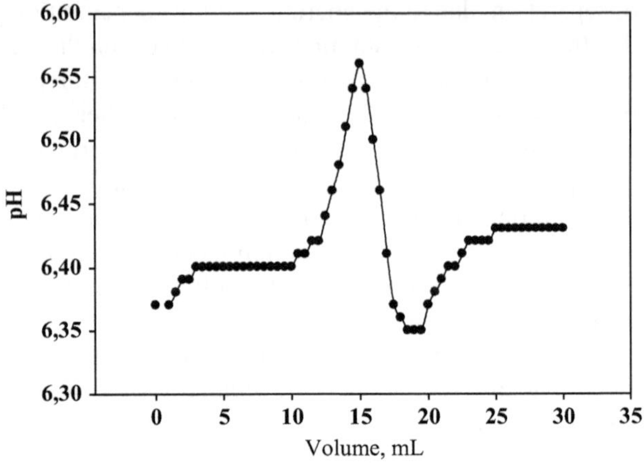

Fig. 2.18 Resolution of tetrabuthylammonium iodide on Styrosorb 2. Column: 28 mL; probe: 3 mL of a 1 % solution; flow rate: 1.2 mL/min of the 0.015 M KCl background electrolyte

In order to achieve an opposite elution order, having the basic fractions before the acidic fractions (Fig. 2.18), a neutral salt composed of large tetrabutylammonium cations and small iodide anions (9.88 and 6.62 Å in diameter, respectively) was injected. In this case a typical hypercrosslinked polystyrene material, Styrosorb 2, which is known to have only small nanopores, was used as the sorbent. Indeed, as expected, the large organic cations were totally excluded from the polymer phase, while the I^- anions were in position to enter certain pores of the packing. Hence, the basic fractions emerged from the column ahead of the acidic fractions.

Naturally, such resolution of a neutral salt into constituent parent acid and base implies the formation of equivalent amounts of new protons and hydroxyls. Although the response of a pH electrode to the concentrations of these two ions is exponential, and the surface area under the negative and positive peaks in the chromatograms cannot be directly correlated with the amounts of the emerging acid and base (the deviation of the base line from pH 7.0 must also be taken into consideration), the rather symmetrical shape of the elution peaks in Figs. 2.17 and 2.18 suggests that the amounts of the acid and base are nearly equivalent.

Of course, generation of additional protons and hydroxyls in the system implies dissociation of an equivalent amount of water. This process proceeds spontaneously even at room temperature, but requires a very substantial amount of energy (118 kcal/mole at 20 °C). This energy can be taken from the kinetic energy of molecules of the liquid phase, which, however, must result in a certain decrease of temperature within the moving chromatographic zone of the salt. Although no quantitative measurements have been carried out as yet, we see a conclusive indication to this mechanism in the fact that increasing the column temperature will result in an increase of the deviations of the effluent's pH from neutrality. At 60 °C these deviations were observed to rise to the magnitude of more than one pH unit in each of the two directions.

2.12 Do we Really Need Sorbent Functional Groups to Separate Electrolytes?

Long-term discussion about the role of functional groups in "acid retardation" process, separation of an acid and salt mixture into individual components on a strong basic anion exchange resin under conditions excluding ion exchange, provoked us to express our opinion of this process. We are convinced that the separation of acids and salts on anion exchange resins also proceeds via size exclusion mechanism, while the contribution of functional groups is miniscule, if any at all. We should only compare the chromatograms of separating the model $CaCl_2$-HCl mixture on the neutral hypercrosslinked NN-381 resin (Fig. 2.2), activated carbon D 4609 (Fig. 2.3) and strong basic anion exchanger PSA-433 (Purolite Int.) taken in Cl^- form (Fig. 2.19) under identical experimental conditions. As can be seen, the profiles of all the chromatograms are identical. The anion exchanger PCA-433 separates well the model mixture, Δ_1=0,31, self-concentration of the salt amounts to 1.23 whereas the concentration of separated hydrochloric acid increases by a factor of 1.59. In this experiment, the breakthrough of hydrochloric acid is observed at $BtV_{0.5}$=25 mL which is practically equal to the column volume, 26 mL, i.e. more than the volume of the mobile phase in the column, 21 mL. Nevertheless, this acid retardation by no means relates to its interaction with quaternary amino groups or styrene-divinylbenzene matrix. Indeed, Fig. 2.20 depicts the sorption isotherm for hydrochloric acid onto PCA-433 in Cl^- form. As can be seen, two hour contact of the exchanger dry beads with the acid solutions of different concentrations results in some HCl retention only if its concentration exceeds 5 mol/L. The high concentrated aqueous acid solutions contain an insufficient amount of water and so hydrogen chloride might already exist as an ionic pair or even in the form of covalent molecules. Dispersion interactions of such HCl with the exchanger can provide its substantial retention. However at the concentrations lower than 5 N, HCl molecules

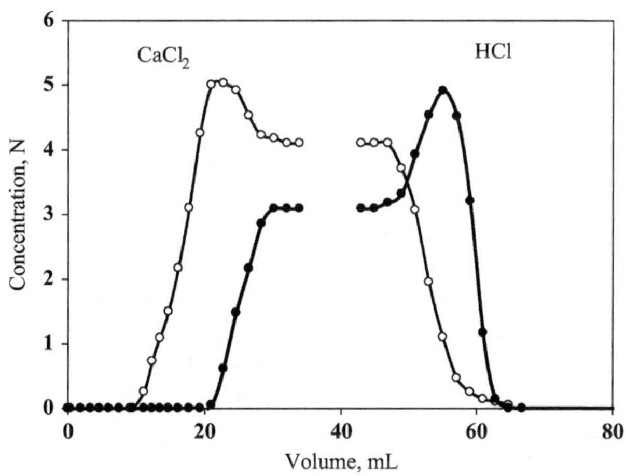

Fig. 2.19 Elution profiles for a $CaCl_2$/HCl mixture from a chromatographic column with strong anion exchange resin PCA-433 taken in Cl^- form

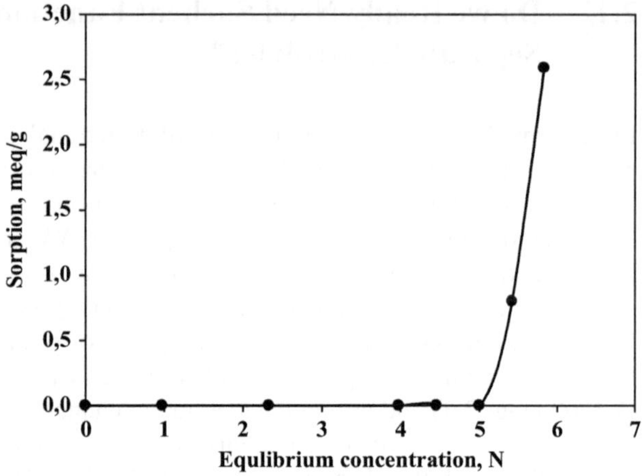

Fig. 2.20 Sorption isotherm for hydrochloric acid onto PCA-433 in Cl^- form

appear to be largely dissociated and display no "over-equivalent" affinity to the anion exchanger in its Cl^- form. Note, that the concentration of mineral acids used on an industrial scale for metal surface etching or metal extraction and processed according to "acid retardation" technique, does not exceed, as a rule, 2 N.

Recently, Khamizov et al. (2012) reported the separation of concentrated mixtures of phosphoric acid and its salts of rare-earth elements via acid retardation process on an anion exchange resin. They also observed an enhanced concentration of the salts and the acid in separated fractions and explained the phenomenon by the *sorption* of both the salts (which in concentrated solutions exist in the form of hydrated ionic pairs) and phosphoric acid (which was assumed to exist in undissociated molecular form). Yet, the authors neither specified the feed concentration, nor proved their suggestions concerning the mechanism of sorption. One wonders why during 50 years that passed since the first publications on acid retardation process, nobody tried to confirm by direct independent experiments the assumption of *sorption* (retention) of acids on anion exchangers in their acidic form.

A question arises why gel-type anion exchange resins having a polystyrene-type matrix crosslinked with 8–10% divinylbenzene happen to be the optimal packing to separate electrolytes. The size of spaces between polymeric chains in the water-swollen polymer network, when calculated for its crosslinking density, would have to be incomparably larger than that of hydrated ions. However, one should take into account that the chloromethylation of a starting copolymer (an intermediate stage in manufacturing anion exchangers) is accompanied by a side reaction of chloromethyl groups with phenyl rings of other polystyrene chains (Anderson 1964). The logical consequence of the side reaction consists of an uncontrolled increase in the real crosslinking density of the starting copolymer and, hence, the decrease in size of its "pores". (Actually, synthesis of hypercrosslinked polystyrene networks capitalizes on the use of the same chloromethylation and intentional post-crosslinking reactions). This explanation is in good agreement with the fact that the gel-type cation exchange resin PCR-833 (Purolite Int.) also containing 8–10% divinylbenzene, does not separate a mixture of 1.2 N Na_2SO_4 and 1.4 N NaCl, since sulfonation of the same starting copolymer does not cause formation of additional crosslinks. At the same time, hypercrosslinked polystyrene NN-381 resin separates the above mixture with a good selectivity of $\Delta_1 = 0.20$.

The essential role of tertiary amine functional groups of the anion exchanger used in the acid-retardation process is thus reduced to providing appropriate expansion of the network through swelling in aqueous media, the property that in hypercrosslinked networks results automatically from the rigidity of its open network structure. A significant shortcoming of conventional gel-type anion exchange resins, due to the presence of above ionogenic groups and flexibility of the network, consists in that their beads shrink when water is replaced with concentrated electrolyte solutions and vice versa. This "breathing" of sorbent beads reduces the lifespan of the material and complicates considerably technical execution of the separation process. Distinctly, neutral hypercrosslinked sorbents maintain their constant volume at any pH and any ionic strength of solutions. Furthermore, anion exchangers can separate only mixtures of electrolytes with common anions, i.e. acids and their salts. In contrast, non-functionalized hypercrosslinked sorbents are universal, they are not endangered by ion exchange phenomena and can differentiate any mixtures of acids, salts and/or bases.

2.13 Productivity of the Ion Size Exclusion Process

The self-concentrating effect definitely enhances the practical value and the total productivity of the ion size exclusion process. Still, for estimating quantitatively its contribution, it is advisable to find concentration C_{av} graphically averaged over the total span Δ of the fractions containing the isolated component, and then the averaged self-concentrating coefficients C_{av}/C_0 for each of the two electrolytes (see Fig. 2.3). A product *P* of the Δ values with the corresponding averaged self-concentrating coefficients will then characterize the total productivity of the separation process; it presents (expressed in bed volumes) the volume of the initial mixture, from which the corresponding electrolyte can be isolated:

$$P = \Delta \times \frac{C_{av}}{C_0}$$

Notably, contrary to the selectivity values Δ and the self-concentrating coefficients, which usually are different for the two components, the corresponding process productivity *P* values are nearly equal. Such values can be found in Table 2.5, showing that the column productivity *P* can easily amount to 0.3–0.5 bed volumes for the comparable and not too high concentrations of two electrolytes. In extreme cases, more than 5–8 bed volumes of the feed solution can be processed in one run, if traces of small ions (such as Cl^- or NH_4^+) are to be removed from very concentrated brines.

Table 2.5 Separation of electrolytes on hypercrosslinked polystyrene resins (Davankov and Tsyurupa 2005)

Electrolyte	MN-270						MN-202						MN-500					
	C_0 N	$BtV_{0.05}$ mL	$BtV_{0.5}$ mL	C_{max}/C_0	Δ_1 BV	P BV	C_0 N	$BtV_{0.5}$ mL	$BtV_{0.5}$ mL	C_{max}/C_0	Δ_1 BV	P BV	C_0 N	$BtV_{0.05}$ mL	$BtV_{0.5}$ mL	C_{max}/C_0	Δ_1 BV	P BV
$CaCl_2$	3.6	15	22.5	1.34	0.25	0.29	3.5	18	23.4	1.14	0.20	0.22						
HCl	3.8	28	30.0	1.50	0.25	0.31	3.9	27.5	29.5	1.41	0.15	0.18						
LiCl	2.1	16.2	20.9	1.08	0.24	0.25	3.85	19.5	23.5	1.15	0.18	0.19						
HCl	2.1	25.8	28.0	1.36	0.20	0.24	4.0	26,3	29	1.45	0.14	0.17						
KCl	1.2	14	17,3	1.08	0.19	0.20												
HCl	4.2	21	22.9	1.08	0.17	0.18												
H_2SO_4	4.0	17	22.4	1.07	0.14	0.14	4.0	20.5	24.9	1.05	0.06	0.06	4.5	18	23.2	1.00	0.05	0.05
HCl	3.9	23	26.5	1.12	0.05	0.05	3.9	23.5	26.8	1.07	0.05	0.05	3.5	19	24.6	1.06	0	
Na_2SO_4	1.8	13.5	17.0	1.17	0.32	0.35												
NaOH	0.95	23	26.5	1.15														
$(Fe+Cu)SO_4$	2.15	13	18.0		0.28	0.28												
H_2SO_4	2.75	19	26.4															
$Al_2(SO_4)_3$	0.9	13	18.0		0.26	0.26												
H_2SO_4	3.4	18	25.7															
$CaCl_2$							1.8		23.4	1.01	0.05	0.05	4.0	19.5	23	1.00	0.05	0.05
$Ca(NO_3)_2$							1.7		24.9	1.06	0.08	0.08	4.3	19.5	24.5	1.06	0.05	0.05
NaCl							1.9		23.3	1.0	0	0						
NaOH							2.2		23.3	1.0	0	0						

C_0 Initial electrolyte concentration,
C_{max} Concentration at maximum of breakthrough curve,
$BtV_{0.5}$ Breakthrough volume at the middle of front,
$\Delta = (Bt_1V_{0.5} - Bt_2V_{0.5})/30$ = Selectivity of separation, 30 mL being the volume of column,
$P = \Delta \cdot C_{av}$ Productivity of separation

2.14 Ion Size Exclusion—Green Technology

When considering the practical aspects of the ion size exclusion process some feasible modes of its realization should be compared.

We usually apply the technique originally developed for the "acid retardation" process, namely, loading the column with the feed solution from the bottom up, and then eluting its content with water pumped in the opposite direction. This approach helped to reveal all details and potentials of the ion size exclusion process. However, from typical chromatograms such as the one presented in Fig. 2.2, one can immediately note that the middle part of the chromatogram, where the concentrations in all effluent fractions are equal to those in the feed, is not productive. Indeed, after determining the process productivity by such type of preliminary experiments, this part of the chromatogram can be easily minimized to no more than one bed volume; onc only has to stop introducing the feed solution (the direct experiment) in the moment that the first eluting component becomes contaminated by traces of the later emerging electrolyte. This would produce a fraction with the first target product composed of larger ions. The mixed intermediate fraction obtained at the beginning of the elution step with water (reversed experiment) can then be returned to the feed reservoir. Finally, third fraction can be collected containing relatively pure target product composed of later eluting smaller ions. Clearly, the process can be easily automated.

Another approach for handling automatically electrolyte mixtures is a pulse-type introduction of feed solutions and water in a constant direction, as in experiments by Hatch and Dillon (Hatch and Dillon 1963). Depending on the ratio of the sampling and washing volumes, a more or less complete separation of two components can be obtained. In this process chromatographic zones have to be automatically cut into fractions of desired concentrations and purity. In order to achieve high flow rates and high productivity of this mode, better quality of column packing is needed, and this can be provided by using smaller bead particles with a narrower bead size distribution. In general, reducing particle size is the common approach to enhance the performance of any chromatographic process.

The best way of separating mixtures of two electrolytes in a continuous procedure is provided by the simulated moving bed (SMB) technique. It uses a series of packed columns connected into a cycle (or carousel) by multi-port valves. This method permits a continuous introduction of the feed solution and the eluent (water) into connecting tubes between the columns and a continuous withdrawal of the two separated fractions from two other connection points (Fig. 2.21). The input and withdrawal positions are changed along a circle according to an optimized program for opening and closing the valves. Results of a mathematical simulation of the ion size exclusion process in the SMB mode have recently been published (Laatikainen et al. 2007). There, phase distribution coefficients obtained for a model $CaCl_2$/HCl mixture were used. This model predicts full feasibility of a continuous SMB process that, for the first time, should permit to obtain both components with concentrations higher than in the initial mixture, while securing the high purity of both products. Indeed, the SMB process has become a mature large-scale technology in resolving

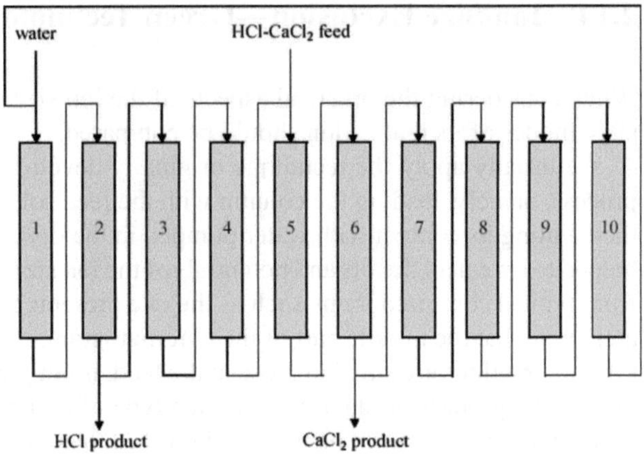

Fig. 2.21 Configuration of the model SMB system

racemic compounds into constituent enantiomers, as it is especially effective in the separation of binary mixtures. However, by involving a more complex scheme of valve switching, continuous separation of three-component mixtures is also possible (Müller-Späth et al. 2008).

The most obvious immediate advantage of the ion size exclusion mechanism is that the elution process of electrolytes from the column does not require any displacing reagent, any acid or base. Pure water can easily carry all the non-retained ions through the column, so that the latter does not need any regeneration after the water wash. For this reason no waste streams are generated, which, in the case of regeneration of ion exchange columns, represent highly mineralized solutions that are expensive to dispose.

The "green" concept of ion size exclusion thus sets new goals and new criteria for the evaluation of preparative chromatographic separations.

2.15 Conclusion

A new method of separation of concentrated electrolyte solutions into individual components by means of preparative frontal ion size exclusion chromatography on neutral nanoporous hypercrosslinked sorbents undoubtedly is of great theoretical interest. The exclusion chromatography is widely used only for the determination of molecular weight and molecular weight distribution of polymers which are injected into column as analytical probe and then eluted with mobile phase. Until we began this study, nobody even tried to use the exclusion chromatography in frontal mode, all the more, for separation of the simplest inorganic electrolytes. This new approach permits to find out new phenomena which were unknown before. First of all, this is the striking spontaneous increase in the concentration of separated electrolytes.

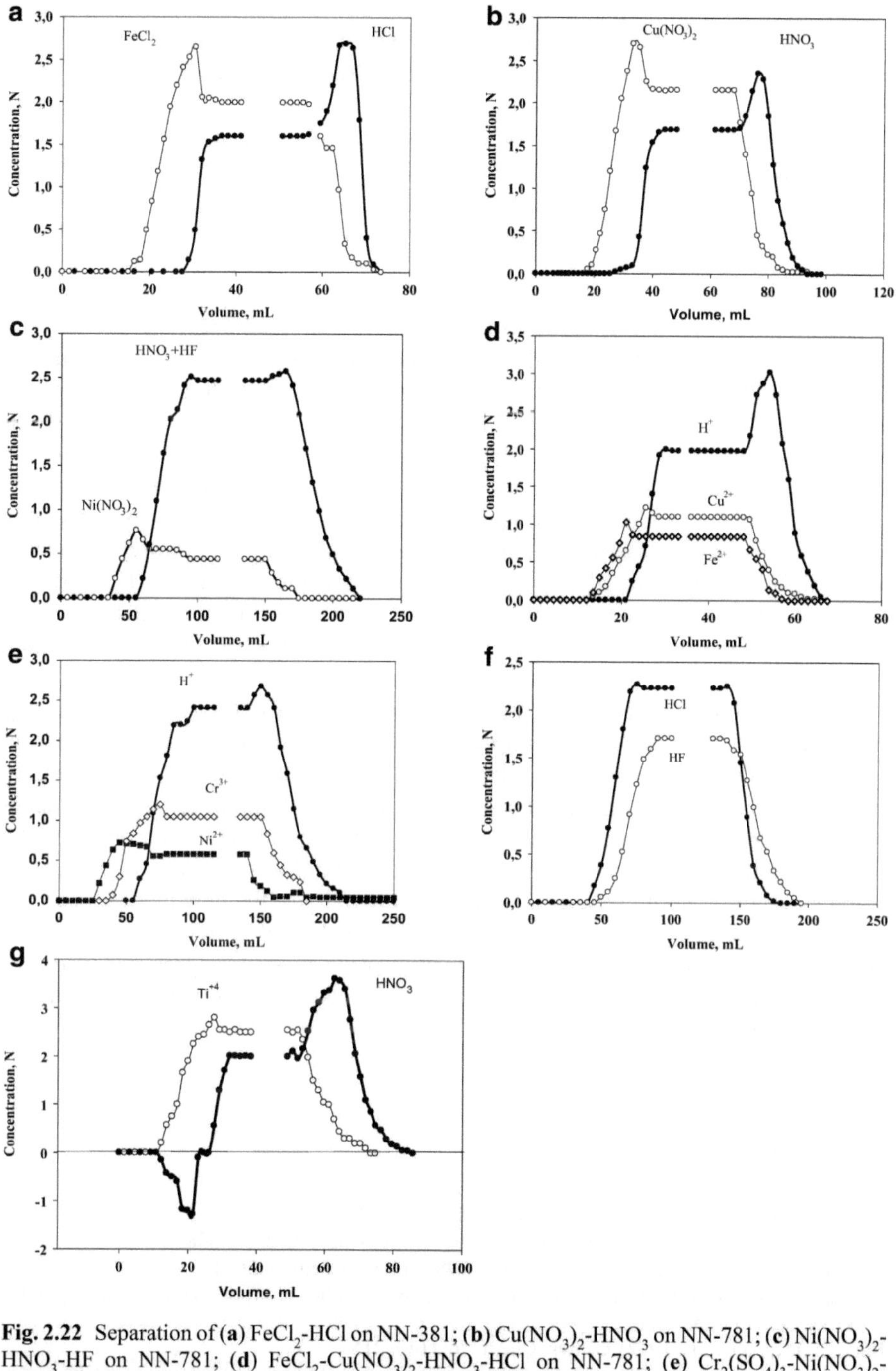

Fig. 2.22 Separation of (**a**) $FeCl_2$-HCl on NN-381; (**b**) $Cu(NO_3)_2$-HNO_3 on NN-781; (**c**) $Ni(NO_3)_2$-HNO_3-HF on NN-781; (**d**) $FeCl_2$-$Cu(NO_3)_2$-HNO_3-HCl on NN-781; (**e**) $Cr_2(SO_4)_3$-$Ni(NO_3)_2$-HNO_3-HF on NN-781, (**f**) HCl-HF on NN-381; (**g**) $Ti(NO_3)_4$-HNO_3 on NN-781

The method quickly established itself as a productive preparative technique suitable for industrial application. Potentials of the method are much greater than those of the acid retardation process on an anion exchange resin that allowed only separation of an acid and its salt. When using hypercrosslinked polystyrene sorbents, one can separate effectively any combinations of acids, salts and bases, including multi-component mixtures. The only essential precondition for the separation is that target ions must differ in their size. Some examples demonstrating the potential uses of NN-381 and NN-781 hypercrosslinked resins in these processes are depicted in Fig. 2.22.

Especially interesting and industrially important could be separation of $Ti(NO_3)_4$-HNO_3 mixture (Fig. 2.22g). Due to the large difference in sizes of hydrated Ti^{4+} and NO_3^- ions, high selectivity of separation of the two components is achieved (Δ_1 amounts to 0.39 and Δ_2 to 0.40). The remarkable fact in this example is that the titanium(VI) salt undergoes spontaneous hydrolysis during its separation from excess acid and converts into TiO(II) salt. Two acid molecules are liberated in this moment, which again separate from the salt. The process resembles resolution of a salt into parent acid and base described in Figs. 2.17 and 2.18 of this review. In this way, more acid can be recovered than was present in the initial mixture in the form of free acid. Correspondingly, less alkali is now needed to precipitate TiO_2, the actual target of the separation process, from the first metal-containing fractions of the effluent. Self-hydrolysis of titanium(IV)-salts (or salts of other polyvalent metals) greatly enhances the economic efficiency of the new process.

Contrary to ion exchange, the suggested method does not require usage of any chemicals besides water and therefore produces no mineralized wastes. Since in the exclusion mechanism of separation there exist no equivalent or complementary retentive sorbent-sorbate interactions, such fundamental notion as sorption capacity does not operate anymore. Therefore, there are no principle limitations as to the concentration of electrolyte solutions to be processed. As opposed to all chromatographic practices, both the selectivity and productivity of the new method increase with the increase in concentration of electrolyte solutions. Chemical passivity of hypercrosslinked sorbents permits their application to such aggressive media as HNO_3, HF, H_2SO_4, which is very important for hydrometallurgical, electrochemical, electronic and other branches of industry.

References

Anderson RE (1964) Ind Eng Chem Prod Res Develop 3:85–89

Bergstrom PA, Lindgren J, Kristiansson O (1991) J Phys Chem 95:8575–8580

Biggin S, Enderby JE, Hahn RL, Narten AH (1984) J Phys Chem 88:3634–3638

Blinnikova ZK, Maerle KV, Tsyurupa MP, Davankov VA (2009) Sorption and Chromatographic Processes (Russian) 9:323–331

Bocris JOM, Saluya P (1972) J. Phys. Chem. 76:2140–2151

Brown CJ, Fletcher CJ (1988). In: Streat M, Ion Exchange for Industry, Ellis Horwood, Chichester, pp 392–403

Brown CJ, Sheedy V, Palaologou M, Thompson R (1997) Proceedings of Annual Meeting of Minerals, Metals, Materials Society, Orlando, USA, TP126
Davankov V, Tsyurupa M (2005) J Chromatogr A 1087:3–12
Davankov V, Tsyurupa M (2006) J Chromatogr A 1136:118–122
Davankov V, Tsyurupa M, Ilyin M, Pavlova L (2002) J Chromatogr A 965:65–73
Davankov VA, Rogozhin SV, Tsyurupa MP (1969) Pat. USSR 299165; Pat. USA 3729457; Chem. Abstr., 75 (1971) 6841B
Davankov VA, Tsyurupa MP (1990) React Polym 13:27–42
Davankov VA, Tsyurupa MP (2011) Hypercrosslinked Polymer Networks and Adsorbong Materials, Elsevier, vol 56
Davankov VA, Tsyurupa MP, Alexienko NN (2005) Mendeleev Commun N 5:192–193
Davankov VA, Tsyurupa MP, Blinnikova ZK (2009) J Sep Sci 32:64–73
Davankov VA, Tsyurupa MP, Alexienko NN (2005) J Chromatogr A 1100:32–39
Dow Chemical Co., Midland, Mich., Tech. Service Bull. 164–62, "Ion Retardation"
Fabricand BP, Goldenberg S (1961) J Chem Phys 34:1624–1628
Ferapontov NB, Gorshkov VI, Parbuzina LR, Trobov HT, Strusovskaya NL (1999) React Funct Polym 41:213–225
Götzelmann W, Hartinger L, Gülbas M (1987) Teil 1, Metalloberfläche, 41:208–212; Teil 2, Metalloberfläche, 41:315–322
Guioshon G (2005) J Chromatogr A 1079:7–23
Hatch JA, Dillon HB (1963) Ind Eng Chem Process Design and Development 2(4):253–263
Hatch JA, Dillon HB, Smith HB (1957) Ind Eng Chem 49:1812–1819
Helfferich F, Exchange I (1962) Ion Exchange. McGraw-Hill, NY, p 134
Hribar B, Southall NT, Vlachy V, Dill KA (2002) J Am Chem Soc 124:12302–12311
Khamizov RKh, Krachak AN, Gruzdeva AN, Bolotokov AA, Khamizov SKh, Smirnov AA, Zhiguleva TI (2012) Sorption and Chromatographic Processes (Russian) 12:29–39
Laatikainen M, Sainio T, Davankov V, Tsyurupa M, Blinnikova Z, Paatero E (2007) J Chromatogr A 1149:245–253
Laatikainen M, Sainio T, Davankov V, Tsyurupa M, Blinnikova Z, Paatero E (2007) React Funct Polym 67:1589–1598
Licheri G, Piccaluga G, Pinna G (1976) J Chem Phys 64:2437–2441
Ma JC, Dougherty DA (1997) Chem Rev 97:1303–1324
Macronet Hypersol, Purolite Int., UK
Manalo GD, Turse R, Rieman WM III (1959) Anal Chim Acta 21:383–391
Marcus Y, Ion Solvation (1985) Wiley, New York
Mecozzi S, West AP Jr, Dougherty DA (1996) J Am Chem Soc 118:2307–2308
Müller-Späth M, Aumann L, Melter L, Ströhlein G, Morbidelli M (2008) Biotechnology and Bioengineering 100:1166–1177
Nelson F, Kraus KA (1958) J Am Chem Soc 80:4154–4161
Nightingale ER (1959) J Phys Chem 63:1381
Pinto NG, Graham EE (2004) Multicomponent Diffusion in Concentrated Electrolyte Solutions: Effect of Solvation. AIChE Journal 33:436
Preparative Enantioselective Chromatography (2005) ed. Cox G.B. Blackwell Publishing, Oxford
Proskurina NA, Il'in MM, Davankov VA, Sychev KS, Kostikov SY (2007) Russian J Physical Chem 81:424–427
Samoilov OY (1972) in "Water and Aqueous Solutions: Structure, Thermodynamics, and Transport Processes". In Horne RA (ed) Wiley-Interscience, New York, pp 597–612
Samoilov OY (1957) Structure of Electrolyte Solutions and Hydration of Ions (Russian), Izdatelstvo AN SSSR, Moskva Samoil
Soldatov VS, Polhovsky EM, Sosinovich ZI (2004) React Funct Polym 60:41–48
Soper AK, Weckström K (2006) Biophys Chem 124:180–191
Sychov CS, Davankov VA, Proskurina NA, Mikheeva AJu (2009) LC-GC Europe 22(1):20–27
Sychov CS, Ilyin MM, Davankov VA, Sochilina KO (2004) J Chromatogr A 1030:17–24

Tanganov BB (2005) Chemical Methods of Analysis (Russian), Izdatelstvo VSGTU, Ulan-Ude, p 550
Tsyurupa MP, Blinnikova ZK, Pavlova LA, Davankov VA (2008) In "Recent Advances in Ion Exchange. Theory and Practice". Proceedings of IEX 2008 (Cambridge), ed. M. Cox, SCI, 77–84
Tsyurupa MP, Blinnirova ZK, Davankov VA (2006) Sorption and Chromatographic Processes (Russian) 6:878–883
Tsyurupa MP, Davankov VA, Doklady Akad Nauk RAN (2004) 398:198–200
Tsyurupa MP, Maslova LA, Andreeva AI, Mrachkovskaya TA, Davankov VA (1995) React Polym 25(1):69–78
Tsyurupa MP, Tarabaeva OG, Pastukhov AV, Davankov VA (2003) Intern J Polym Materials 52:403–414
Wheaton RM, Bauman WC (1953) Ind Eng Chem 45:228–234

Chapter 3
Supercritical Fluid Chromatography: A Green Approach for Separation and Purification of Organic and Inorganic Analytes

Sriparna Datta, Runa Ghosh Auddy and Amit De

Abstract 'Supercritical fluid' describes a gas or liquid at conditions above its critical point. A greater range of solvent properties can be achieved with Supercritical fluid as a single solvent by careful manipulation of temperature and pressure at the supercritical state. Supercritical fluids are attractive media for several chemical reactions having better control over the reaction rates in different areas of biochemistry, polymer chemistry and environmental science. Supercritical fluid extraction (SFE), a rapid, convenient, efficient, and selective method has been used successfully for the separation of analytes prior to supercritical fluid chromatography (SFC), which is a relatively recent chromatographic technique and is commercially available since 1982. SFC significantly reduces the usage of organic solvents and wastes by using supercritical CO_2 as the mobile phase. The important principles of green chemistry that are applicable to green chromatography includes prevention of waste, use of safer solvents and increasing energy efficiency. All these factors are taken care of in SFC which combines some of the best features of HPLC as well as GC. Analytes that cannot be vaporized for analysis by gas chromatography or have no functional groups for detection by the usual liquid chromatography techniques, can be separated and detected using SFC. By now SFC has been applied to wide variety of materials including natural products, pharmaceuticals, foods, pesticides, herbicides, surfactants, polymers and polymer additives, heavy metals, fossils fuels, petroleum, explosives and propellants. SFC has now become an attractive alternative for chiral drug separation.

S. Datta (✉) · R. G. Auddy · A. De
Department of Chemical Technology, University of Calcutta, 92 A. P. C Road, Kolkata 700009 West Bengal, India
e-mail: sriparna_d@yahoo.com

R. G. Auddy
e-mail: auddyruna@gmail.com

Inamuddin, A. Mohammad (eds.), *Green Chromatographic Techniques*,
DOI 10.1007/978-94-007-7735-4_3, © Springer Science+Business Media Dordrecht 2014

3.1 Introduction to Green Chemistry and Supercritical Fluid Chromatography

L. H. Lawrence, defined green chemistry as *"the use of analytical chemistry techniques and methodologies that reduce or eliminate solvents, reagents, preservatives and other chemicals that are hazardous to human health or the environment and that may also enable faster and more energy efficient analyses, without compromising performance criteria"*. It encompasses all those areas which utilize natural processes i.e., biosynthesis, biocatalysts, biochemical engineering and biotechnology-based transformations, judicious use of eco friendly solvents, use of innocuous reagents, development of renewable feedstocks, use of microwave assisted synthesis and design of chemical transformations that reduce the use of required energy (Tundo and Anastas 2000). Green analytical methods utilize supercritical fluids, micro-wave treatments and ultra-sound instead of liquid solvents and adhere to the set of 12 principles of green chemistry proposed in 1998 by Anastas (Anastas and Warner 1998) for extraction and separation of analytes. Green analytical chemistry utilizes newer eco-friendly technologies and modifies old approaches to perform rapid analytical activities. By these techniques a variety of surfactants, plant bioactives, antibiotics, polycyclic aromatic hydrocarbons and other organic compounds have been extracted without/minimal use of solvents, thus minimizing the negative impact on the environment.

Gas chromatography(GC) can be regarded green since the gasses used viz. hydrogen and helium are non-polluting but the sample preparation for GC requires solvents which are harmful to the environment. The green approach for sample preparation in GC is primarily based on miniaturization of procedures and the use of solvent-less techniques. In the next phase, purging inert gas through the solution and trapping analytes on a sorbent reduces pollution. Thermal desorption of pollutants collected on the sorbent is generally used for measurement of volatile organic compounds in environmental air. The process of thermal desorption meets the requirements of green chemistry for gas chromatographic analysis (Wardencki and Namieśnik 2002). Liquid Chromatography (LC) mainly high performance liquid chromatography (HPLC) generates a large amount of organic toxic solvents. Acetonitrile (ACN) is one of the commonly used organic solvents but its disposal is a major concern to chromatographers. Thus, reducing acetonitrile is one of the first options in greening LC. According to the green solvent selection guide (Alfonsi and Colberg 2008), water, acetone, ethanol and methanol can be considered environmentally friendly for LC applications. Reduction in column size, diameter, particle size, using elevated temperature, changing pH of the eluent, using hydrophilic interaction chromatography (HILIC) is preferred for the separation of polar and ionizable compounds, all of which conforms to green chromatography (Alpert 1900). At such elevated temperatures, stationary phase can be synthesized by temperature-responsive polymer with reversible hydrophilic-hydrophobic conformation for modulation of retention. Such stationary phases also contribute to green chromatography. Polymers like poly (N-isopropylacrylamide) (PIPAAm) have been used with only aqueous mobile phase for separation of bioactive compounds. In HILIC, separation is achieved by partitioning between a water-enriched layer on the surface of a polar stationary phase and a mobile phase containing a high percentage of an organic solvent. Ethanol mixed with CO_2 has been introduced in HILIC as a green alternative (Gama et al. 2012). Reversing the mobile phase constituents in HILIC with high concentrations of water and low concentrations (or zero) of an organic solvent is another green alternative for the analysis of ionizable and polar compounds. This technique is called per-aqueous liquid chromatography (PALC) (Periera et al. 2009). Chromatographic separation of niacin and niacinamide using pure water as the sole component of the mobile phase has been investigated. The major advantage is the total elimination of organic solvents as required in HPLC (Yang et al. 2011).

The main principle of green technology is to minimize the amount of reagents consumed and toxic wastes generated. Automation is the best strategy available because of enhanced sampling frequency and integration of all the steps of analytical methods including on-line treatment of wastes (Armenta et al. 2008). Taking into account all the factors like savings in energy and consumables, adaptation of mechanized procedures and also elimination of waste, reduction of treatment costs, green analytical chemistry is an economic and environment-friendly process.

Supercritical fluid chromatography (SFC) has been one of the greatest breakthroughs in analytical chemistry in recent times (Taylor 2009). Analysis of compounds begins with sample preparation, separation, detection and data handling. The applicability of supercritical fluid is not only related to chromatography but also to extraction (Charpentier and Sevenants 1988), chemical reactions (Jessop et al. 1995) and remediation of environment (Sako et al. 1997).

3.2 Super Critical Fluids

The term '*supercritical fluid*' describes a gas or liquid at conditions above its critical point. Every substance has a critical point on the temperature and pressure curve at which the difference between its liquid and gaseous state disappears and the compound is in its supercritical state. Slight changes in temperature and pressure around the critical point causes significant changes in density and other physical properties of the fluid. *Super critical fluid* (*SCF*) can be described as a fluid obtained by heating—a compound above the critical temperature and compressing above the critical pressure. SCFs have a pressure-tunable dissolving power (Wang et al. 2003). They also have liquid-like densities and gas-like flow properties. Thus a greater range of solvent properties can be achieved with a single solvent by careful manipulation of temperature and pressure at the supercritical state. Therefore using solvents in their supercritical condition expands their solubility, volatility and polarity, which can be utilized to develop new analytical methods. There is no surface tension of a supercritical fluid, as there is no liquid/gas phase boundary.

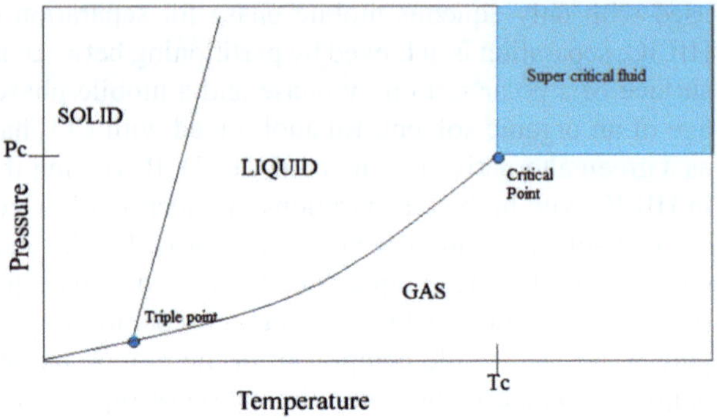

Fig. 3.1 Phase diagram for pure substance

In order to illustrate; carbon dioxide (CO_2) can exist in three phases—solid, liquid and gas depending on various temperature and pressure conditions. For CO_2, like any other substance there is a temperature above which it cannot remain liquid even at very high pressure. Accordingly, there is a pressure above which it cannot remain as a gas even on increasing the temperature. This temperature is known as critical temperature (T_c) and the pressure is known as critical pressure (P_c) and the state is known as *Critical Point*. At this stage liquid and gas phase of a substance co-exists and both have the same density. Above this point no phase change is observed and the fluid is designated as *Super Critical* (Fig. 3.1).

It is evident from this phase diagram that a gas on increasing temperature above the critical point behaves as a super critical fluid and similarly by increasing pressure above the critical pressure, the liquid behaves as super critical fluid. Therefore, any compound, mixture or element can behave as a super critical fluid above its critical temperature or pressure. In case of solid phase, the molecules of a substance remain closely placed following a specific geometry. In the liquid phase, the distance between the molecules increases and in gaseous phase the molecules are randomly distributed. Above critical point, the distribution of the molecules is intermediate between liquid and gases as represented below. The density, viscosity and diffusion co-efficient of SCF are intermediate between gases and liquids.

Solubility of a material in a fluid is the most important property to be considered which can be manipulated by slight change in temperatures and pressures. Solubility in a supercritical fluid at a constant temperature tends to increase with density of the fluid (Lu et al. 1990). Since density increases with pressure, solubility tends to increase with pressure. Similar relationship with temperature is more complicated. At constant density, solubility increases with temperature. However, near the critical point, the density drops sharply with a slight increase in temperature hence solubility also drops with increasing temperature, near critical temperature and then again increases. Advantages of SCFs are; they are inexpensive, innocuous, eco-friendly and non-toxic. SCFs have the advantage of higher diffusion constants and lower viscosities relative to common liquid solvents. These two properties become important in chromatographic analysis and extractions with SCFs. Low viscosity means reduced pressure drop across the column for a given flow rate.

Table 3.1 Critical properties of some commonly used SCFs

Fluid	Critical temperature (°K)	Critical pressure (bar)
Carbon dioxide	304.1	73.8
Ethane	305.4	48.8
Ethylene	282.4	50.4
Propane	369.8	42.5
Propylene	364.9	46.0
Trimethoflurane	299.3	48.6
Chlorotrifluoromethane	302.0	38.7
Trichloromethane	471.2	44.1
Ammonia	405.5	113.5
Water	647.3	221.2
Cyclohexane	553.5	40.7
n-Pentane	469.7	33.7
Toluene	591.8	41.0

Higher diffusion coefficient gives higher analysis speed. SCFs have high densities (0.2–0.5gm/cm^3) due to which they have a remarkable ability to dissolve large, non-volatile molecules. Dissolved analytes can be easily recovered by simply allowing the solutions to equilibrate with the atmosphere at low temperatures thus proving useful for unstable analytes. The critical properties of some commonly used supercritical fluids are listed in Table 3.1.

Certain disadvantages of SCFs are; increased development time and multiple factors for control and hence the methods using SCFs are not well developed like GC and HPLC.

However, the unique combination of properties of SFCs are ideally suited for developing processes for extracting, purifying, and recrystallizing fine chemicals and pharmaceuticals. SCFs are also finding applications in fractionation of low vapour pressure oils, in several reactions in different areas of biochemistry, polymer chemistry, inorganic and organic chemistry, environmental sciences as well as food, polymer and material industries (Greibrokk 1995).

The two supercritical fluids of particular interest are carbon dioxide and water. Carbon dioxide is a non-flammable, non-toxic, recoverable and eco-friendly solvent with low critical temperature of 304 °K and moderate critical pressure of 73 bar. Water at supercritical conditions not only dissolves ionic species at ambient conditions but also dissolves paraffin, aromatics, gases and salts. Due to this unique property, supercritical water can be used in separation processes of toxic wastewater. The choice of SCFs depends on their properties, safety, cost and the specific application.

3.2.1 Supercritical Fluid Extraction (SFE)

Chromatography and extraction are two closely related analytical processes used extensively for chemical separation and isolation of compounds (King et al. 1993). The most commonly used separating phases are gases and liquids but with the emergence of green technologies, supercritical fluids have gained acceptance as alternate methods to solvent extraction. Compared with conventional extraction, SFE is relatively fast because of the greater solute- fluid binary diffusion coefficient than liquid -liquid systems, and their use minimizes the problem of solvent waste disposal. Supercritical fluids can adjust their solubilising property primarily via the process of compression and temperature adjustment thereby enabling the extraction of one or more analytes of varying polarities and solubility using one supercritical fluid. Use of supercritical fluid improves the extraction kinetics. The SFE technique has been used to extract organic substances such as lipids and hydrocarbons, polymer additives, pesticides and chlorinated compounds in environmental samples, bioactives from natural products and determination of drugs in body fluids and tissues. Other applications include determination of explosives and propellants for forensic investigations (Wang et al. 2003). Chronologically, analytical SFE developed later than SFC (King et al. 1993) but today it can be applied for extraction of samples (nanogram levels) and coupled with chromatographic methods such as GC, LFC, SFC and MS. The supercritical fluid of choice is CO_2 because it is suitable for extraction of thermally labile compounds and is a non-toxic extraction medium, thus minimizing the cost and potential hazards associated with solvents (Lang and Wai 2001). The extraction also involves the mass transfer of the extracted solute into the supercritical fluid. This is directly proportional to solubility of the solute in the supercritical fluid and the diffusivity.

The kinetics of extraction in supercritical fluid is similar to liquid-liquid extraction (Fig. 3.2). This can be pictorially represented as follows:

The extraction process is theoretically presented as

$$\frac{m}{m_0} = \left(\frac{6}{\pi^2}\right) \sum_{n=1}^{\infty} \left(\frac{1}{n^2}\right) exp\left(\frac{-n\pi^2 Dt}{r^2}\right)$$

Where, m is the mass of the extractable material remaining in the sample

m_0 is the mass of the original extractable material
n is an integer
D is diffusion co-efficient in the matrix of radius r
t denotes the time for extraction.

This equation can be rewritten in terms of reduced time $t_r = \left(\frac{\pi^2 Dt}{r^2}\right)$ and the equation becomes

$$\frac{m}{m_0} = 6\pi^2 \left[exp(-t_r) + exp(4t_r) + exp(-9t_r) + \ldots\ldots\ldots\right]$$

which explains the extraction process followed in SFE.

The extraction process in SFE can be enhanced by the addition of modifiers (Taylor and King 1993). Addition of methanol as modifier with CO_2 not only enhances the solubilisation of polar compounds, but increases the solubility of lipophilic solute by over 100 folds.

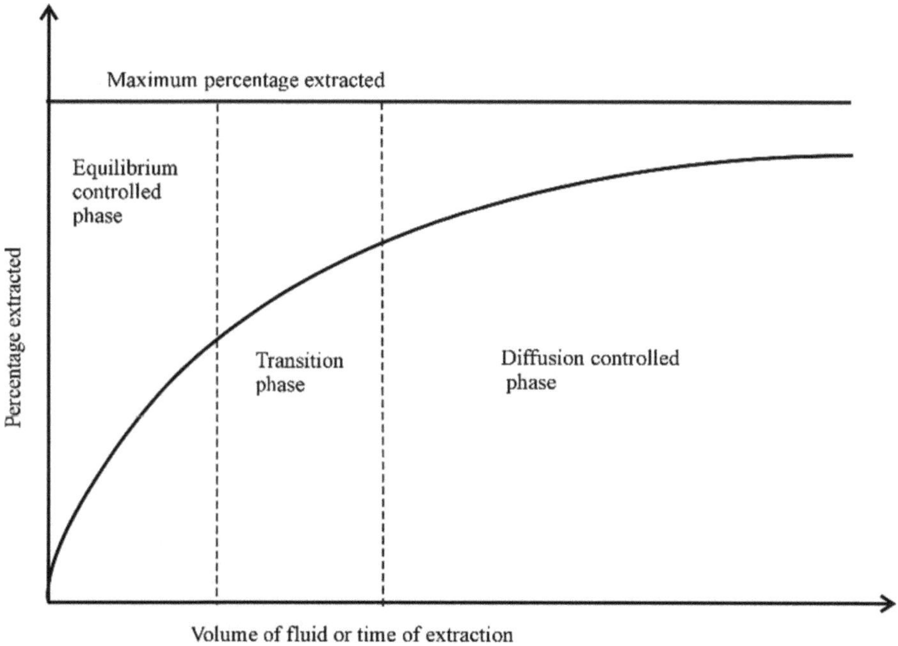

Fig. 3.2 Generalized extraction curve of any solute against percentage of extracted material and the volume of fluid used for extraction or time of extraction

The general procedure for such extraction is as follows; the supercritical fluid is delivered by a pump into an extraction cell where the extraction takes place under controlled pressure and temperature conditions. Then the extract is swept into an injection system for chromatographic analysis. Analytical SFE is usually measured in an off-line or on-line mode, the latter technique involves the direct transfer of the extract into a chromatographic column. Off-line mode permits separation of large sample sizes whereas the online mode is suitable for characterizing small samples and is more sensitive (King et al. 1993).

In an off-line system, the extraction fluid is delivered through a pump, whose heads are cooled to a sub-ambient temperature to assure liquefaction of the gas. A second pumping module is also connected in-line to add a modifier for the extraction fluid. The extraction vessel is contained in a heater assembly that equilibrates the extraction fluid to the desired temperature. After extraction, the fluid pressure is reduced through a back pressure regulator, and the analytes of interest are collected in a vial containing a solvent, cooled by the adiabatic, iso-enthalpic expansion of the expanded fluid stream. A pictorial presentation of the layout is presented in Fig. 3.3.

In on-line mode (Fig. 3.4), a single pump is used to deliver the fluid for both the extraction and the chromatographic steps. A multiport switching valve in conjunction with the column shut-off valve permits the diversion of the fluid either into the extraction cell or the column. Extract from the cell is subsequently trapped in a T-shaped junction before commencement of the chromatography step. This is ac-

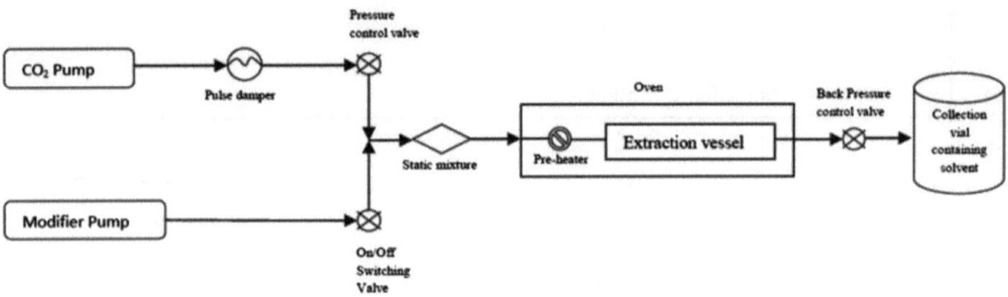

Fig. 3.3 Layout of a typical offline supercritical fluid extraction system

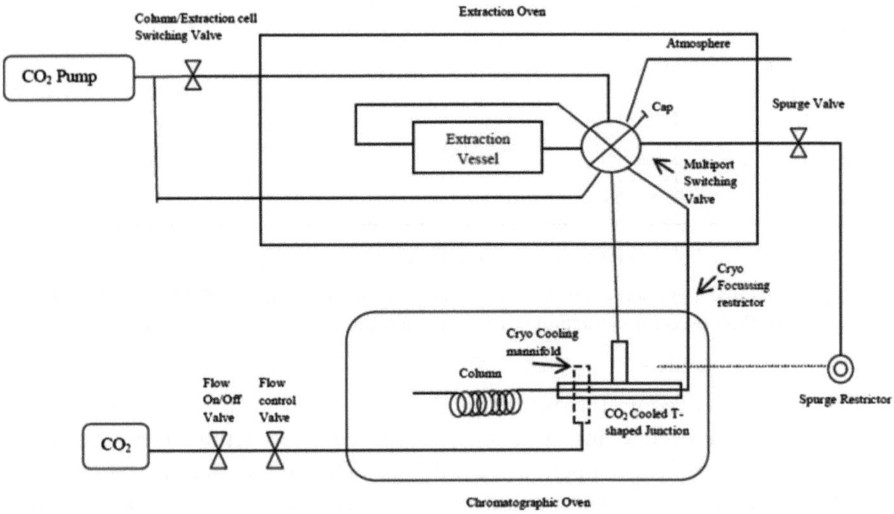

Fig. 3.4 On-line system for supercritical fluid extraction

complished by re-switching the valve so as to effect desorption of the sample from the CO_2 cooled T-shaped junction onto the head of the column. This layout allows the purging of the extraction vessel during the chromatographic analysis.

The effect of variables such as temperature, pressure, type of supercritical fluid, nature of modifiers and the solubility parameters of the analyte have considerable impact on the results obtained by SFE (Yamini et al. 2002). Other factors are the type of collection techniques employed and the influence of the sample matrix on the efficiency of extraction (King et al. 1993). Increasing pressure of the supercritical fluid at constant temperature results in increasing density and there by higher solvent power of the supercritical fluid (Turner et al. 2001). The solubility characteristics of a supercritical fluid are directly related to its density (King 1989). The relationship was used by Giddings and co-workers (Giddings et al. 1969) to design a solubility parameter scale. When the density of the supercritical fluid is equal to the target analyte, maximum solubility is achieved in the super critical fluid. CO_2, at high densities can have solvent properties similar to liquids like chloroform and acetone. The flow rate is also an important parameter which is to be decided on the basis of sample size and kinetics of extraction. Collection techniques(Turner et al. 2002) applied for SFE involves total depressurization of the extract into a suitable collection device, use of adsorbent columns or cryo-cooling, partial depressurization of the extract into a reservoir or sample loop, with subsequent recycling of the extraction fluid back to the extraction cell, ion exchange columns etc. Of these, the most popular collection method involves total depressurization of the extract into a suitable collection device.

Several sample matrix parameters influence the end result that is obtained by SFE (Yamini et al. 2002). These parameters are particle size and shape, surface area and porosity, moisture content, sample size and the level of extractability.

3.3 Supercritical Fluid Chromatography (SFC): An Overview

Chromatography is based on a flow system containing two phases, one mobile and the other stationary. The underlying principle involved in chromatography is the distribution of an analyte between two phases, a separating phase and a stationary phase. The separating phase is called the mobile phase and the stationary phase is an immobilized liquid or solid phase over which the mobile phase passes.

Quantitatively the distribution of an analyte between two phases can be expressed as $K = C_m/C_s$, where K is called the partition coefficient, C_m represents the concentration of the analyte in the mobile (or extracting) phase, and C_s represents the concentration of the analyte in the stationary phase. In extraction, this distribution is used to separate the analyte from the sample. In chromatography, compounds with different K values can be isolated from each other through repetitive distributions between a separating (mobile) phase passing through a stationary phase. Among the various chromatographic techniques, the simplified mechanism is; attaining equilibrium between the stationary and the mobile phases and distribution of the substance under examination between these two phases whereby separation can be achieved. SFC has emerged as one of the most recent green chromatographic techniques used in the modern era of science and technology. SFC is an eco-friendly alternative to conventional solvent extraction methods such as liquid chromatography and a viable alternative to gas chromatography for volatile and semi-volatile compounds.

3.3.1 History of Development of Supercritical Fluid Chromatography

The term, supercritical chromatography with mobile phases in the supercritical state was first suggested by James E Lovelock in 1958 (Sethi et al. 2010). The first experiments were however performed by Ernst Klesper et al. (1962), on the separation and purification of thermo-labile porphyrin derivatives using supercritical chlorofluoromethane at pressures up to 140 bar and temperatures from 150 °C to 170 °C (Klesper et al. 1962). This method was further developed theoretically as well as experimentally by other workers in 1960s. At the Pittsburgh Conference 1981–1982, Hewlett-Packard introduced SFC instrumentation for packed columns where Dennis R. Gere and his associates at Hewlett-Packard modified a commercial HPLC instrument for supercritical fluid chromatography (Gere et al. 1988). They used 10 cm × 4.6 mm i.d. columns packed with 3/μm particles of a bonded-phase material and CO_2 as mobile phase (Turner et al. 2001). Milos Novotny and Lee developed capillary SFC in 1982 and the first commercial capillary column instrument was introduced in 1985 using both CO_2 and n-pentane as the mobile phases (Novotny et al. 1981). This type of SFC was regarded as an extension of GC. Capillary SFC combined supercritical mobile phases and open tubular fused silica column technology. However, in the 90s, packed column SFC was introduced which utilized the same injector and packed column configurations as in HPLC. It is more useful for routine separation of a wide range of pharmaceuticals and is more convenient than the open tubular capillary column. With packed columns, binary and ternary components in the mobile phase can be used along with composition programming. Non-destructive detectors like UV-detectors, PDA detectors became compatible with this type of system. Packed columns in SFC is usually operated near the critical temperature of the fluid with flow control pumps and electronically controlled back pressure regulators mounted downstream of the column to maintain optimal flow rates and mobile phase compositions. This assures proper volumetric mixing of the main fluid and the modifier for gradient elution.

In the last two decades, supercritical fluids and their application in chromatography gained significant interest for their unique properties and low toxicity to the environment. SFC can be applied for separation/analysis of a wide range of compounds which include organic compounds, ecdysteroids, plant bioactives, acidic drugs and basic drugs, pesticides in soil, chiral drug enantiomers and vitamins.

3.3.2 Instrumentation

The instrumentation includes pump, column, injector port, column oven and detector (Fig. 3.5). Normally in SFC the fluid used as eluent is usually compressed above its critical pressure in order to attain the corresponding liquid state. It is pumped into the column and before entering the column it is heated above its critical temperature (Kirchner and Taylor 1993). Inside the column a constant pressure is maintained using the pressure restrictor present on the downstream side of the column or after the detector. The flow diagram is presented in Fig. 3.5.

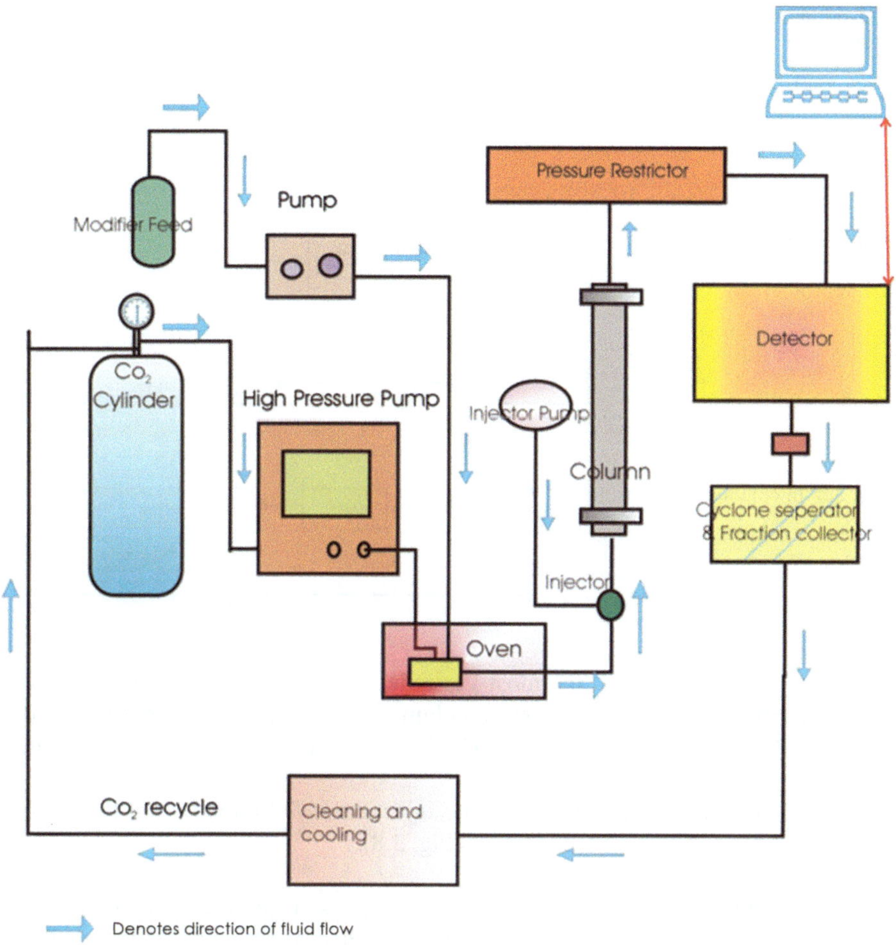

Fig. 3.5 Pictorial layout of SFC system

Pumps High pressure pumps are required for the operation of SFC systems in order to maintain a constant pressure and a pulse less flow of the eluent. The choice of pump depends on the type of column used for analysis. For packed column SFCs usually a reciprocating pump is used. In case of open tubular column, a syringe pump is usually preferred. An oven is also used for the precise control of temperature of the mobile phase. These ovens are similar to convention LC and GC ovens.

Injectors The injection in an SFC system can be carried out in three procedures (Berger 1995; Majewski et al. 2005).

1. Loop injection: This type of injection system requires a low pressure pump to fill the loop. This type of injection is usually carried out for preliminary tests of column performance and elution parameters (Fig. 3.6). The pictorial layout is presented as follows:

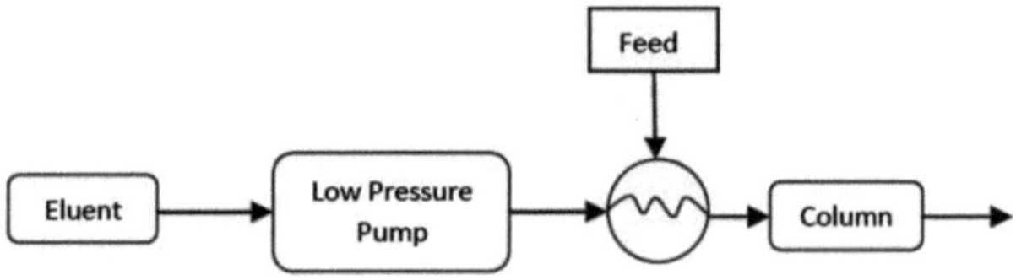

Fig. 3.6 Loop injection system

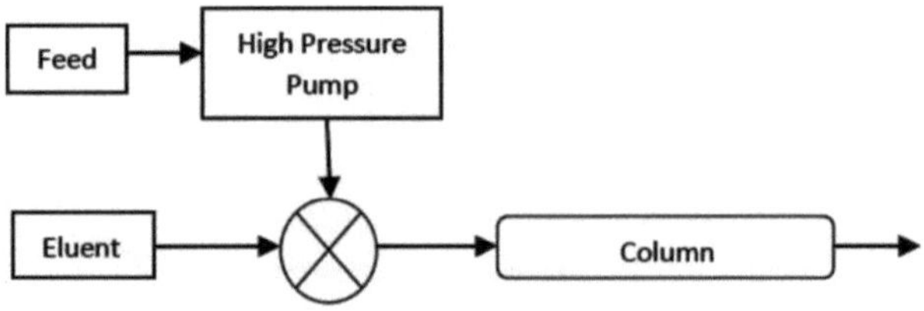

Fig. 3.7 In line injection

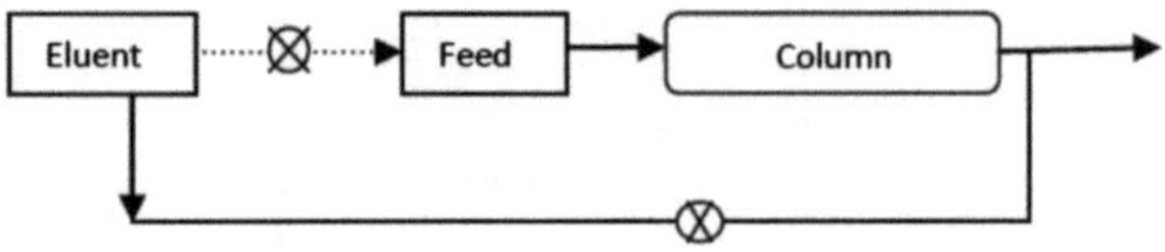

Fig. 3.8 In column injection

2. In line injection: This is the most versatile mode of injection as it has a better flexibility of changing the injection volume. A high pressure pump is required to inject the feed solution (Fig. 3.7). The injected stream is dissolved in the eluent flow.
3. In column injection: In this case the injection is fed directly into the column without any dilution with the eluent (Fig. 3.8).

Injector volumes In an SFC system the mobile phase is pumped in its liquid state. Before entering the column, the mobile phase is heated above its critical temperature. Just before entering the column, the sample is injected into the eluent stream through an injection and finally into the column. A restrictor present at the downstream of the column or after the detector maintains a constant pressure required to maintain the supercritical condition of the fluid within the column. Sometimes heating is required to prevent the clogging of the restrictor. Both variables and fixed restrictors are used for the process. The injection volume for packed column is kept above 1 μL and for open tubular column it is about 96 nL. A greater volume affects resolution between the eluting peaks.

Types of columns used for SFC:

a. Open tubular columns: In this case wall coated glass columns having very small internal diameter varying between 50–500 μm are usually used. The column efficiency is very high in these columns, however the same decrease with the increase of density of the fluid as in this case the diffusivity also reduced significantly. With the rise in temperature diffusivity, solubility and volatility of liquid increased. A minimal pressure drop effect is also observed with this type of column. The plate height (H) with this type of column is usually represented as:

$$H = \frac{2D_m}{u} + \frac{d_c{}^2\left(1 + 6k + 11k^2\right)u}{96(1 + k)^2 D_m} + \frac{2kd_t{}^2u}{3(1 + k)^2 D_s}$$

Where,

D_m = Mobile phase solute diffusion coefficient
d_c = Column diameter
k = Capacity factor
d_t = Stationary phase film thickness
u = Linear velocity of mobile phase
D_s = Stationary phase solute diffusion coefficient

b. Packed columns: These columns are most popular and used in modern day SFC systems. In packed columns the permeability decreases and the flow resistance increases. So, the column length is restricted to 25 cms for packed columns. The efficiency in packed column increases with the increase in particle density and decrease in particle size. The plate height decreases with the decrease in particle size. Increasing the particle size decreases the pressure drop, increases the permeability and decreases the solubility.

$$H = 1.5d_p + \frac{1.4D_m}{u} + \frac{2(k_0+k+k_0k)^2 d_p{}^2 u}{15(1 + k_0)^2(1 + k)^2 D_m}$$

Where,

H = Plate height
D_m = Mobile phase solute diffusion coefficient
d_p = Particle diameter
k = Capacity factor
u = Linear velocity of mobile phase
k_0 = Ratio of particle pore velocity to particle interstitial velocity (typically 0.5)

The retention time (T_r) is represented as a function of column length, linear velocity of the mobile phase and capacity factor as follows

$$T_r = \frac{L}{u}(1 + k)$$

Packed columns are usually suitable for higher density analytes. Packed capillary columns are suitable for mass flow detectors; a greater sensitivity is obtained. Due to larger column length, a higher permeability is obtained which makes it suitable for use with multiple detector systems. The different column dimensions used in SFC are tabulated in Table 3.2.

Stationary phase The stationary phase used in columns are quiet similar to those used with GC and HPLC. The basic packing material is either silica or alumina which is useful for the analysis of non-polar compounds but not suitable for polar

Table 3.2 Column dimensions used in SFC

Column type	Length (cm)	Internal diameter (mm)
Conventional packed column	3–25	2–4.6
Microbore packed column	3–25	0.5–2
Packed capillary column	5–50	0.1–0.5
Open tubular column	100–3500	0.025–0.1

solutes because of the possibility of irreversible adsorption in some cases. To reduce polarity of the stationary phase it is often bonded to long carbon chains like octyl, octadecyl, cyanoalkyl, aminoalkyl, dialkyl groups (King et al. 1993). This increases the permeability but requires a polar modifier along with the supercritical fluid as eluent. Apart from these, the widely used polar stationary phases are polysiloxanes which may be substituted depending on the type of analysis. Substituted polysiloxanes includes polymethyl siloxanes suitable for the separation of closely eluting polar components or cyanopropyl polysiloxanes sutaible for the separation of components with carboxylic groups.

Mobile phase Carbon dioxide is the most preferred fluid for use in SFC. The critical temperature T_c for carbon dioxide being 31 °C is very low compared to several other fluids that can be used as supercritical fluid. It is also comparatively less toxic and environment friendly.

A polar modifier is often required because its addition increases the solubility of the polar analyte by altering the polarity of the eluent. It reduces the retention volume, and the strong interaction between the absorptive site and the polar analyte. However, the choice of the modifier must be such that it remains miscible with the super critical fluid over the entire range of the temperature and pressure of analysis. The different types of modifiers that can be used are tabulated in Table 3.3.

Detectors Various types of detectors can be used with SFC systems. These includes open cell detectors i.e. detectors used in conventional gas chromatographs with slight modifications and closed cell detectors like those used in conventional HPLC systems.

The open cell detectors producing ionization under ambient pressure include:

1. Flame ionization detector (FID) used in the analysis of organics
2. Hydrogen atmosphere FID for organometallics
3. Thermionic detectors for nitrogen and phosphorus containing compounds
4. Electron capture detectors (ECD) for the analysis of halogenated or electronegative compounds
5. Photo-ionization detectors for organics with ionization potentials less than 10.2 eV.
6. Ion mobility detectors for high electron/proton affinity compounds

The closed cell detectors include

Table 3.3 Different types of modifiers

Modifier	Critical temperature T_c (°C)	Critical pressure P_c (atm)	Molecular mass	Dielectric constant at 20 °C	Polarity index
Methanol	239.4	79.9	32.04	32.70	5.1
Ethanol	243.0	63.0	46.07	24.3	4.3
1-Propanol	263.5	51.0	60.10	20.33	4.0
2-Propanol	235.1	47.0	60.10	20.33	4.0
1-Hexanol	336.8	40.0	102.18	13.30	3.5
2-Methoxy ethanol	302.0	52.2	76.10	16.93	5.5
Tetrahydrofuran	267.0	51.2	72.11	7.58	4.0
1,4-Dioxane	314.0	51.4	88.11	2.25	4.8
Acetonitrile	275.0	47.7	41.05	37.50	5.8
Dichloromethane	237.0	60.0	84.93	8.93	
Chloroform	263.2	54.2	119.38	4.81	4.1
Propylene carbonate	352.0		102.09	69.0	6.1
N, N-Dimethyl acetamide	384.0		87.12	37.78	6.5
Dimethyl sulfoxide	465.0		78.13	46.68	7.2
Formic acid	307.0		46.02	58.50	
Water	374.1	217.6	18.01	80.10	10.2
Carbon disulfide	279.0	78.0	76.13	2.64	

1. Optical Detectors:
    a. UV –Detector for compounds having chromophoric groups
    b. FT-IR
    c. Flame photometric detectors for sulphur and phosphorus containing compounds
    d. Chemiluminescence detectors for sulphur and easily oxidizable compounds
    e. Element selective plasma emission detectors
    f. Light scattering detectors
2. Mass Detectors

3.3.2.1 Advantages and Disadvantages of Supercritical Fluid Chromatography

Advantages:

1. Provides rapid separations without the use of organic solvents. Thus it is considered as a green chromatographic technique.
2. Faster separation process reduces the time for analysis achieved through a decrease in resistance to mass transfer in column.
3. High resolution at lower temperature can be achieved
4. Low viscosity and high diffusion coefficient

Disadvantages:

1. Constant maintenance of pressure and temperature makes the process expensive.
2. Cleaning is time consuming.

3.3.3 Properties of SFC compared to GC and HPLC

Super Critical Fluid Chromatography is one of the most recent and revolutionary separation techniques used in modern science and technology. The use of supercritical fluid as the mobile phase makes it different from both GC and HPLC. It can be considered as hybrid of GC and HPLC since the SCF above its critical temperature and below its critical pressure is a gas whereas when it is below critical temperature and above critical pressure it is a liquid. SCF has densities and solvating power like liquids but lower viscosity and high diffusivity and hence it can act as solvent carriers like the mobile phase of GC as well as solvent for these substances like liquid in HPLC (Taylor 2009). The initial publicity talked about the SFC of having all the advantages of GC and HPLC but none of its disadvantages (Smith 1999). SFC offers advantages over high performance liquid chromatography (HPLC), mainly because SF shows high efficiency of mass transfer and affords fast speed and fair resolution in routine analysis, particularly in a complex mixture. SFC may be considered to be a separation technique intermediate between HPLC and GC. In SFC the selectivity, permeability, solubility and plate count, all three are intermediate between GC and HPLC. SFC is compatible with both HPLC and GC detectors. Multi-detector system compatible with SFC makes it universally applicable for large number of analytes (Sethi et al. 2010). Lower viscosity, higher diffusivity of SCFs compared to the solvents of HPLC allow lower pressure drop in analytical columns. Thus increase in column length, along with faster flow rates can be achieved which in turn affect capacity ratios, selectivity and theoretical plate heights. However, SFC has not yet come into use as a total substitute for GC or HPLC in the separation science community (Majewski et al. 2005; Smith et al. 1988; Kirschner and Taylor 1993).

3.4 Industrial Applications of SCFs and SFCs

Since 1970s industrial applications of SCFs were being explored and in 1980's, supercritical carbon dioxide became the most preferred fluid in SFC. SFC has found application mainly in the analytical field. It has been lamented in general that supercritical fluids are associated with high processing costs which is not true since decaffeinated coffee with supercritical fluid is available at competitive market prices, which can surely contradict this impression. Supercritical fluid chromatography though known from decades, has now become a very useful analytical tool with the supercritical fluid as mobile phase. SFC significantly reduces the usage of organic solvents and waste by using supercritical CO_2 as the mobile phase. Commercial instruments are available and conventional HPLC can also be converted to accommodate SFC. Supercritical CO_2 has very low viscosity and easily diffused causing very fast separation which is even faster than HPLC. CO_2 is nonpolar, but polarity can be increased by doping with methanol in less than 30% concentration. SFC has been successfully used to separate oligomers and chiral compounds (White and Burnett 2005) and high molecular weight polymers like proteins, drugs and steroids (Berger and Wilson 1994a, b; Loran and Cromie 1990). SFC can be coupled with the common detectors including mass spectrometers.

Common areas of application include the following:-

Pharmaceutical Industry Supercritical carbon dioxide can dissolve a broad range of both polar and non-polar compounds and thus has found several applications in the pharmaceutical industry (Sekhon 2010). Supercritical fluid chromatography is being increasingly used to separate and purify active substances from undesired by-products. Particle formation processes using SFC can make the drugs bioavailable by tailoring the particle morphology and reducing the particle size of the drug. Production of micro and nano-particles for drug delivery can be addressed by supercritical fluids. Analytical chiral separations, purification of medical polymers, are some of the other uses of SFC. Impregnation of active compounds into polymers is a developing area under SFC.

It has been successfully used for extraction and purification of vitamin E from palm oil (Lau et al. 2006). Solvent residues in pharmaceutical and food products are being prohibited by the regulatory agencies since 1970s. Residual solvents can be extracted from pharmaceutical preparations by SFCs. Nonvolatile, thermally and chemically labile drug molecules can be safely analyzed by SFC. Estimation of Prostaglandins (Koski and Jansson 1991), combination of non-steroidal anti-inflammatory drugs (Jagota and Stewart 1992), anticancer drugs (Musser and Callery 1990), antipsychotics (Berger and Wilson 1994), vasodilators (Bhoir et al. 1998) and estimation of several other drugs from blood can be done by standard SFC procedures.

Supercritical fluid technology offers a new technique allowing a single-step generation of active pharmaceutical ingredients (API) which is difficult or impossible to obtain by traditional techniques. The generation of pure and dried new co-crystals of API can be achieved utilizing unique properties of SCFs: supercritical CO_2 solvent power, anti-solvent effect and its atomization enhancement (Padrela et al.2009). Recrystallization and preparation of nanosize pharmaceuticals is possible for some pharmaceutical compounds by gas anti-solvent (GAS) process. If a compound is not soluble in a gas, the gas can be used as an anti-solvent, causing re-crystallization by mixing the gas and an organic solvent solution. By manipulating the rapidity and intimacy of mixing, the GAS recrystallization can achieve narrow particle size distribution in the nanometer range (Yeo and Kiran 2005).

Waxy compounds with low melting points are difficult to micronize by conventional milling processes. Supercritical fluid can be employed as a solvent to dissolve these compounds and then by pressure decrease, precipitation in fine size can be realized (1 to 2 micron size).

Liquid and solid surfactants suffer from undesirable color, odour and other impurities and hence cannot find pharmaceutical applications. Supercritical carbon dioxide can purify heat sensitive materials which cannot be processed by distillation or very low vapor pressure. Thus supercritical fluid extraction has the ability to purify the surfactants of importance for pharmaceutical formulations.

Another emerging application of supercritical fluids is the production of nanoparticle formation. Amphotercin B has been impregnated with poly(l-lactic acid) and poly(methyl methacrylate) to form nanoscale particles using a supercritical carbon-dioxide cosolvent method for solubilization and processing (Pathak et al. 2007).

SC-CO_2 at high pressures has antimicrobial properties (Cinquemani and Boyle 2007). Recent studies have proved that SC-CO_2 is an effective alternative for terminal sterilization of biological materials and medical devices (White et al. 2006). The low interfacial tension, viscosity and solvating properties help in the removal of contaminants.

SFC has been successfully utilized for the separation of single ingredients from mixtures of natural products. Capillary supercritical fluid chromatography coupled with supercritical fluid extraction can be used to separate and isolate both moderately polar and non-polar substances with wide molecular weight range (100–1000 daltons) from complex mixtures of natural products (King 1990). Decaffeination of coffee and tea, extraction of hop component, extraction of essential oils, separation of phenolic compounds, separation of triterpenes and sterols from natural sources are some of the applications. Supercritical fluid extraction has emerged as an acceptable technique for extraction of bioactives from herbs (Lang and Wai 2001) Supercritical fluid chromatography, can be applied to purify very-low-concentration of Digitalis-like factor type of biomolecules in hypertensive patients with kidney failure and quantify them using flame ionisation detector (Graves et al. 2000).

Forensic Science In addition to drug analysis, the forensic chemists are also engaged in scrutinizing a variety of samples, such as paints, fire residues, post-explosion debris, gunshot residues, glass, soil, ink, fingerprints, for any trace evidence of crime. Allen et al. compared the recovery of morphine from blood using SFE with that of solid-phase extraction (SPE), and found both to be similar (Allen et al. 1999). In 1994, Edder et al. demonstrated the quantitative extraction of opiates in hair (Edder et al. 1994). Illicit preparations of marijuana and hashish contain more than 400 compounds of differing polarities among which the major cannabinoids are tested in forensic laboratories. SFC coupled to an atmospheric pressure chemical ionization mass spectrometer (SFC-APCI-MS) has been successfully used to separate cannabinoids (Backstrom and Cole 1997).

SFE method has been validated and approved by the US environmental protection agency for routine extraction of total recoverable petroleum hydrocarbons (Method 3560 1994). Compared to other existing methods, SFE is a rapid, convenient, efficient, and selective method for detection of explosives in post-explosion residues (Wang et al. 2003). Addition of an appropriate modifier to the CO_2 and choice of different columns have made the analysis of explosives possible by SFC (McAvoy et al. 1999).

Thermal and Nuclear Power Generation Closed-cycle gas turbines with supercritical carbon dioxide operating near 550 °C, have large implications for bulk thermal and nuclear generation of electricity. Very high thermal efficiencies of carbon dioxide above 500 °C and 20 MPa resulted in the increased production of electrical power per unit volume of fuel by 40 percent or more. Supercritical carbon dioxide as an emerging natural refrigerant is being used in domestic heat pumps (Lorentzen 1994). Supercritical water reactors (SCWRs) are promising advanced nuclear systems that offer similar thermal efficiency gains (Kritzer and Dinjus 2001). Compared to average emissions from the existing coal based fleet, up to 35 % reductions in CO_2 and pollutant emissions can be achieved today in commercial supercritical or ultra supercritical new plants, without increase of cost of electricity.

Carbon dioxide can also be used in supercritical cycle nuclear plants, with similar efficiency gains (Dostal and Hejzlar 2007).

Special Applications Recently, it has been shown that supercritical CO_2 can be used with chelating agents to replace organic solvents for i) the extraction of uranium from aqueous solutions produced in the processing of the ore (Dung le and Imai 2006) and ii) the removal of heavy metals from soils and sludges (Babei and Dacera 2005). Conversion of vegetable oil to biodiesel via trans-esterification reaction, by converting to the methyl ester is usually done using methanol and catalysts. This can be now achieved using supercritical methanol without a catalyst. Thus, the washing of product to remove the catalyst is not required. Furthermore, it is easier to design a continuous process to convert other cooking oils to biodiesel (Bunyakiat et al. 2006).

Chemical Reactions Supercritical fluids are attractive media for several chemical reactions due to better control over the reaction rates in different areas of biochemistry, polymer chemistry and environmental science. Temperature and pressure can tune the reaction through preferred pathways, e.g., to improve yield of a particular chiral isomer (Oakes and Clifford 1999). Supercritical water is used as a medium in a combustion reaction to oxidize organic wastes to convert into inert species. One of the big advantages of SCWO is that the combustion efficiency is greater than 99.999 %. Supercritical water oxidation system has excellent process stability and control. Liquid effluents can be discharged without further treatment whereas gaseous effluents can also be discharged to the atmosphere since no detectable quantity of noxious gases is present.

Food Industry Supercritical carbon dioxide extraction can be used to extract thermally labile undesirable food components simultaneously from coffee, spices and herbs retaining the flavour and aroma characteristics of the product which cannot be obtained by the traditional organic solvent extraction processes. Supercritical carbon dioxide has also been used to extract lilac, black pepper (Sankar 1989), essential oils (Tezel et al. 2000), nutmeg, vanilla, basil, ginger, chamomile, cholesterol and coffee. Commercial process for decaffeination of coffee with super critical fluids is being done in Germany (Saldaña et al.2002).

Shukla et al., 1994 have used supercritical CO_2 to produce butter with specific properties such as reduced moisture content, elevated melting points, and lower cholesterol from milk fat (Shukla et al. 1994). Supercritical fluid extraction has been utilized to extract lipids from a variety of meats, including pork and beef (Taylor et al. 1997; Berg et al. 1997). Supercritical fluid extraction can be tailored to the polarity of the solute of interest by changing the temperature and pressure of the extraction process for removal of pesticides residue from foods (Camel 1998). Estimation of these residual pesticides, herbicides and fungicides can be done by SFC (El-Saeid 2003). High value nutraceuticals can be enriched by fractionation through a two-step process consisting of supercritical fluid extraction and supercritical fluid chromatography (King et al. 1996; Montanari et al. 1996).

Environmental Application Supercritical fluid extraction can be used as an alternative technique for soil remediation and activated carbon regeneration. About 99 % organics including PAHs, PCBs (Jentoft and Gouw 1976), DDT and toxophene and highly polar compounds have been successfully extracted out. Commercial process to separate oils from refinery sludge and contaminated soil and extraction of heavy metals has been developed. Separation of heavy components of crude oil can be commercially done by the SFCs. Supercritical water showing the ability to dissolve many nonpolar organic compounds like alkanes and chlorinated biphenyls and several gases have been used as an attractive media for oxidative reactions of waste water streams from chemical, petroleum, textile industries and used for determination of PAHs in soil and solid wastes (Sunarsao and Ismadji 2009).

Material and Polymer Science Packed-column supercritical fluid chromatography (pSFC) with atmospheric pressure chemical ionization (APCI) detector and mass spectrometry (MS) provides a versatile method for the detection and quantification of 20 polymer additives, including common antioxidants, light stabilizers and slip agents (Carrott and Davidson 1998) and the additives can also be extracted from the polyethylene polymer by SFE. The attributes of supercritical fluids, especially CO_2, are extremely attractive in polymer extraction application. Carbon dioxide swells polymers sufficiently to allow the interior volumes to be reached in order to dissolve and extract out the undesired materials and, when the pressure is reduced; all the carbon dioxide is removed.

GC, HPLC and SEC (size exclusion chromatography are not suitable for characterizing the polymers due to temperature instability, detector incompatibility, non-volatility etc. SFC offers great advantages for qualitative and quantitative analysis of the polymers since it can operate at very low temperature and individual oligomers can be detected with proper molecular weight distribution using proper detector (Knowles et al. 1988).

SCF can be applied for preparation of ceramic precursors using supercritical water by using the corresponding metal oxides (Adschiri et al 1992a). Hydrothermal synthesis of metal (hydrous) oxides from their respective salts using supercritical water (Adschiri et al 1992b). Supercritical water is favourable for fine particle formation due to the unique supersaturation properties and high reaction rates (Adschiri et al. 1992; Hayashi and Hakuta 2010). Polycarbosilanes are the most widely studied class of polymer precursors for silicon carbide and can be prepared by supercritical CO_2. Organic microcellular foams were prepared using supercritical fluid processing followed by supercritical drying (Elliott et al. 1992). Extremely small celled (<2 μm) supermicrocellular foams were synthesized by using supercritical fluids, in which the SCF is used as the foaming agent in the parent material (Cha et al. 1992).

Super Critical Fluid Chromatography Analysing Inorganic Analytes SFC is a versatile technique for the separation of metal chelates, transition metals and heavy metals. Organometallic compounds of lead, mercury, and tin are separated by SFC. Lanthanide and actinide complexes are also separated using SFC. Solubility of organometallic compounds can be easily determined using SFC (Khorassani et al. 1997). Direct extraction of Ni^{+2} and Cu^{+2} from an aqueous matrix is also possible via *in-situ* chelation using diethyldithiocarbamate and bis (trifluoroethyl) dithiocarbamate as the ligands (Zúñiga et al. 2009). A method for the extraction of triethyl lead [$TEL^{(+)}$], trimethyl lead [$TML^{(+)}$], and Pb^{2+} from sand was developed using supercritical fluid extraction (modified) and *in situ* complexation with sodium diethyldithiocarbamate (NaDDTC) and was also successfully applied to lead analysis in sand collected from an oil-polluted beach in Chile (Veriansyah et al. 2005). Fabrication of inorganic/organic hybrid composites like silica/perfluoroalkyl methacrylate polymer (PHDFDMA) particles was done using various types of silica by polymerization in supercritical carbon dioxide. Similar to supercritical carbon dioxide, supercritical water is also considered as medium for particle formation whereby mainly the inorganic compounds like oxides are investigated (Hakuta et al. 2003; Cabanas and Poliako 2001; Savage 2009). The internal structure of water changes in the supercritical state. Hydrogen bonding, responsible for the orientation of the molecules towards each other, start to diminish and weaken when water enters the supercritical state (Yamaguchi 1998). As a result, several characteristics of water like the excellent solvation of inorganic compounds and salts, dielectric constant are changed in supercritical state. Corrosion one of the major obstacles in the industrial applications of supercritical water and can lead to severe problems (Kritzer 2004). Since processes running under supercritical conditions, are at comparably low temperatures, hence show less tendency towards corrosion (Eliaz et al. 2003). But removal of these inorganic compounds from supercritical water is essential since inorganic compounds lead to severe problems like catalyst poisoning, scaling and corrosion (Schubert et al. 2010). Supercritical water due to the changes in solvation behavior and solubility for the inorganic compounds offers an option for separation of inorganic compounds in a solid form from the remaining bulk of fluid by the super critical removal process (SCR). The difference between the actual solubility of a compound in a supercritical state and the concentration in the bulk stream is the driving force for such a separation. Super critical removal is a process which is applied as pre-treatment and as post-treatment of water. It is possible to combine the process of SCR with existing technologies like RO or MSF for such treatments (Hodes et al. 2004).

3.5 Conclusion

Supercritical fluid chromatography (SFC) is emerging as a popular technology in the industry for the separation and purification of a variety of compounds from complex matrices. Improvement in the instrumentation and development of chiral stationary phases has made SFC an attractive emerging technique. The green features of SFC offer potential advantages over HPLC and GC. The technique delivers high productivity, high reproducibility and reduced cost compared with HPLC. Supercritical fluid chromatography offers advantages like speed of analysis due to a lower viscosity and higher diffusion coefficients of feed components in the mobile phase allowing separations to be performed. The supercritical fluid commonly used: CO_2 offers advantages such as being cheap, non-flammable and environmentally benign. Though, much research is needed on SFC, the future looks promising where use of supercritical CO_2 and water would become important for analysis of compounds to minimize the use of organic solvents for environmental safety and sustainability.

References

Adschiri T, Kanazawa K, Arai K (1992a) Rapid and continuous hydrothermal crystallization of metal oxide particles in supercritical water. J Amer Ceram Soc 75:1019–1022

Adschiri T, Kanazawa K, Arai K (1992b) Rapid and continuous hydrothermal synthesis of boehmite particles in subcritical and supercritical water. J Amer Ceram Soc 79:2615–2618

Alfonsi K, Colberg J (2008) A study by pfizer global research and development: green chemistry tools to influence a medicinal chemistry and research chemistry based organisation. Green Chem 10:31–36

Allen DL, Scott KS, Oliver JS (1999) Comparison of solid phase extraction and supercritical fluid extraction for the analysis of morphine in whole blood. J Anal Toxicol 23:216–218

Alpert AJ (1990) Hydrophilic interaction chromatography(HILIC), a new method for separation peptides, nucleic acids and other polar solutes. J Chromatogr 499:177–196

Anastas PT, Warner TC (1998) In green chemistry. Theory and practice. Oxford University Press, New York, p 30

Armenta S, Garrigues S, de la Guardia M (2008) Green Analytical chemistry. TrAC Trends Anal Chem 27(6):497–511

Babei S, Dacera DM (2005) Heavy metal removal from contaminated sludge for land application: a review. Waste Manage 26(9):988–1004

Backstrom B, Cole MD (1997) A preliminary study of the analysis of cannabis by supercritical fluid chromatography with atmospheric pressure chemical ionization mass spectroscopic detection. Sci Justice 37(2):91–97

Berg H, Magard M et al (1997) Development of a supercritical fluid extraction method for determination of lipid classes and total fat in meats and its comparison with conventional methods. J Chromatogr A 785:345–352

Berger TA (1995) Packed column Sfc. RSC chromatography monograph. Royal society of Chemistry, UK

Berger TA, Wilson WH (1994a) Separation of drugs by packed column supercritical fluid chromatography.1. Phenothiazine antipsychotics. J Pharm Sci 83(3):281–286

Berger TA, Wilson WH (1994b) Separation of drugs by packed column supercritical fluid chromatography. 2. Antidepressants. J Pharm Sci 83(3):287–290
Bhoir IC, Raman B et al (1998) Separation and estimation seven vasodilators using packed column supercritical fluid chromatography. J Pharm Biomed Anal 17(3):534–546
Bunyakiat K, Makmee S et al (2006) Continuous production of biodiesel via transesterification from vegetable oils in supercritical methanol. Energy Fuels 20(2):812–817
Cabanas A, Poliako M (2001) The continuous hydrothermal synthesis of nano-particulate ferrites in near critical and supercritical water. J Mater Chem 11(5):1408–1416
Camel V (1998) Supercritical fluid extraction as a useful method for pesticides determination. Analysis 26(6):99–111
Carrott MJ, Davidson G (1998) Identification and analysis of polymer additives using packed-column supercritical fluid chromatography with APCI mass spectrometric detection. Analyst 123:1827–1833
Cha et al (1992) Microcellular thermoplastic foamed with supercritical fluid, US Patent 5,158,986 dated Oct 7, 1992
Charpentier BA, Sevenants MR (eds) (1988) Supercritical fluid extraction and chromatography—techniques and applications, ACS symposium series, No. 366. American Chemical Society, Washington, DC
Cinquemani C, Boyle C (2007) Inactivation of microbes using compressed carbon dioxide—An environmentally sound disinfection process of medical fabrics. J Supercrit Fluids 42(3):1–6
Dostal MJ, Hejzlar DP (2007) A supercritical carbon dioxide cycle for next generation nuclear reactors. http://hdl.handle.net/1721.1/17746
Dung le TK, Imai T (2006) Extraction of uranium from simulated ore by the supercritical carbon dioxide fluid extraction method with nitric acid-TBP complex. Anal Sci 22(11):1425–1430
Edder P, Staub C et al (1994) Supercritical fluid extraction of opiates in hair of drug addicts. J Chromatogr B 658:75–86
Eliaz N, Mitton DB, Latanision RM (2003) Review of materials issues in supercritical water oxidation systems and the need for corrosion control. Trans Indian Inst Met 56(3):MIAC4
Elliott J (1992) Microcellular Foams, US Patent 5,182,382. July 7:1992
El-Saeid MH (2003) Pesticide residues in canned foods, fruits and vegetables: the application of supercritical fluid extraction and chromatographic techniques in the analysis. Sci World J 3:1314–1326
Gama MR, Da Costa Silva RG et al (2012) Hydrophilic interaction chromatography. TrAC Trends Anal Chem. Accessed 9 May 2012
Giddings JC, Myers MN, King JW (1969) Dense gas chromatography pressures to 2000 atmospheres. J Chromatogr Sci 7:276–283
Gere DR, Houck RK et al (1988) Supercritical fluid chromatography: a technology update. Fresenius J Anal Chem 330:222–224
Graves SW, Markides KE, Hollenberg NK (2000) Application of supercritical fluid chromatography to characterize a labile digitalis-like factor. Hypertension 36:1059–1064
Greibrokk T (1995) Applications of supercritical fluid extraction in multidimensional systems. J Chromatogr A 703:523–536
Hakuta Y, Haganuma T et al (2003) Continuous production of phosphor yag: Tb nanoparticles by hydrothermal synthesis in supercritical water. Mater Res Bull 38(7):1257–1265
Hayashi H, Hakuta Y (2010) Hydrothermal synthesis of metal oxide nanoparticles in supercritical water. Materials 3:3794–3817. doi:10.3390/ma3073794
Hodes M, Marrone PA et al (2004) Salt precipitation and scale control in supercritical water oxidation. Part A: fundamentals and research. J Supercrit Fluids 29(3):265–288
Jagota NK, Stewart JT (1992) Separation of non-steroidal anti-inflammatory agents using supercritical fluid chromatography. J Chromatogr 604(2):255–260
Jentoft RE, Gouw TH (1976) Analysis of polynuclear aromatic hydrocarbons in automobile exhaust by supercritical fluid chromatography. Anal Chem 48(14):2195–2200
Jessop PG, Ikariya T, Noyori R (1995) Homogeneous catalysis in supercritical fluids. Science 269:1065–1069

Khorassani MA, Combs MT, Taylor LT (1997) Solubility of metal chelates and their extraction from an aqueous environment via supercritical CO_2. Talanta 44(5):755–763
King JW (1989) Fundamentals and applications of supercritical fluid extraction in chromatographic science. J Chromatogr Sci 27:355–364
King JW (1990) Fundamentals and applications of supercritical fluid extraction in chromatographic science. J Chromatogr Sci 28:9–14
King JW, Hill HH Jr, Lee ML (1993) Analytical supercritical fluid chromatography and extraction, Supplement and Cumulative Index. In: Rossiter BW, Baetzold RC (eds) Physical methods of chemistry series, 2nd ed. Wiley
King JW, Favati F, Taylor SL (1996) Production of tocopherol concentrates by supercritical fluid extraction and chromatography. Sep Sci Technol 31:1843–1857
Kirchner CH, Taylor LT (1993) Recent advances in sample introduction for supercritical fluid chromatography. J High Resolut Chromatogr 16:73–84
Klesper E, Corwin AH, Turner DA (1962) High pressure gas chromatography above critical temperatures. J Org Chem 27:700–701
Knowles DE, Nixon L et al (1988) Industrial applications of super critical fluid chromatography polymer analysis. Fresenius J Anal Chem 330(3):225–228
Koski IJ, Jansson BA (1991) Analysis of prostaglandins in aqueous solution by supercritical fluid extraction and chromatography. J Pharm Biomed Anal 9(4):281–290
Kritzer P (2004) Corrosion in high-temperature and supercritical water and aqueous solutions: A review. J Supercrit Fluids 29(1–2):1–29
Kritzer P, Dinjus E (2001) An assessment of supercritical water oxidation (SCWO): Existing problems, possible solutions and new reactor concepts. Chem Eng J 83(3):207–214
Lang Q, Wai CM (2001) Supercritical fluid extraction in herbal and natural product studies—a practical review. Talanta 53:771–782
Lau HLN, Choo Y et al (2006) Quality of residual oil from palm-pressed mesocarp fiber (Elaeis guineensis) using supercritical CO_2 with and without ethanol. J Am Oil Chem Soc 83:893–898
Lawrence LH www.ChemistsHelpingChemists.org
Loran JS, Cromie KD (1990) Evaluation of use of supercritical fluid chromatography with light scattering determination for the analysis of steroids. J Pharm Biomed Anal 8(7):607–671
Lorentzen G (1994) Revival of carbon dioxide as a refrigerant. Int J Refrig 17(5):292–301
Lu BCY, Zhang D, Sheng W (1990) Solubility enhancement in supercritical solvents. Pure Appl Chem 62(12):2277–2285
Majewski W, Valery E, Ludemann-Hombourger O (2005) Principle and application of super critical fluid chromatography. J Liq Chrom Related Technol 28:1233–1252
Method 3560 (1994) Supercritical fluid extraction of total recoverable petroleum hydrocarbons. In: Test methods for evaluating solid waste, physical/chemical Methods (SW-846). Rev.2 United States Environmental Protection Agency
McAvoy Y, Dost K et al (1999) Preliminary study of the analysis of explosives using packed-column supercritical fluid chromatography with atmospheric pressure chemical ionization mass spectrometric detection. Forensic Sci Int 99:123–141
Montanari L, King JW et al (1996) Selective extraction of phospholipid mixtures by supercritical carbon dioxide and cosolvents. J Food Sci 61:1230–1233
Musser SM, Callery PS (1990) Supercritical fluid chromatography/chemical ionization/mass spectroscopy of some anticancer drugs in a thermospray ion source. Biomed Environ Mass Spectrom 19(6):348–352
Novotny M, Springston SR et al (1981) Capillary supercritical fluid chromatography. Anal Chem 53:407A
Oakes RS, Clifford AA (1999) Sulfur oxidation in supercritical carbon dioxide: dramatic pressure dependent enhancement of diastereoselectivity for sulphoxidation of cysteine derivatives. Chem Commun (3):247–248
Padrela L, Rodrigues MA et al (2009) Formation of indomethacin–saccharin cocrystals using supercritical fluid technology. Europ J Pharm Sci 38:9–17

Pathak P, Meziani MJ, Desai T et al (2007) Supercritical fluid processing of drug nanoparticles in stable suspension. J Nanosci Nanotechnol 7:2542–2545
Periera A, David F et al (2009) The acetonitrile shortage: Is reversed HILIC with water an alternative for the analysis of highly polar ionizable solutes? J Sep Sci 32(12):2001–2007
Sako T, Sugeta T et al (1997) Decomposition of dioxins in fly ash with supercritical water oxidation. J Chem Eng Japan 30:744–747
Saldaña MDA, Zetzl C, et al (2002) Decaffeination of guaraná seeds in a microextraction column using water-saturated CO_2. J Supercrit Fluids 22:119–127
Sankar KU (1989) Studies on the physicochemical characteristics of volatile oil from pepper extracted by supercritical carbon dioxide. J Sci Food Agric 49:483–493
Savage PE (2009) A perspective on catalysis in sub- and supercritical water. J Supercrit Fluids 47(3):407–414
Schubert M, Regler JW, Vogel F (2010) Continuous salt precipitation and separation from supercritical water. Part 1: Type 1 salts. J Supercrit Fluids 52(1):99–112
Sethi N, Anand A, Jain G, Srinivas KS, Chandrul KK (2010) Supercritical fluid chromatography—A hybrid of GC and LC. Chron Young Sci 1(2):12–22
Sekhon BS (2010) Supercritical fluid technology: an overview of pharmaceutical applications. Int J PharmTech Res 2(1):810–826
Shukla A, Bhaskar AR et al (1994) Physicochemical and rheological properties of butter made from supercritically fractionated milk-fat. J Dairy Sci 77(1):45–54
Smith RM (1999) Supercritical fluids in separation science—the dreams, the reality, and the future. J Chromatogr 856:83–115
Smith RD, Wright BW, Yonker CR (1988) Supercritical fluid chromatography; current status and prognosis. Anal Chem 60:1323A–1336A
Sunarsao J, Ismadji S (2009) Decontamination of hazardous substances from solid matrices and liquids using supercritical fluids extraction: review. J Hazard Mater 161(2009):1–20
Taylor LT (2009) Supercritical fluid chromatography for the 21st century. J Supercrit Fluids 47:566–573
Taylor SL, King JW (1993) Analytical-scale supercritical fluid extraction of aflatoxin B1 from field-inoculated corn. J Agric Food Chem 41(6):910–913
Taylor SL, Eller FJ, King JW (1997) A comparison of oil and fat content in oilseeds and ground beef- using supercritical fluid extraction and related analytical techniques. Food Res Int 30(5):365–370
Taylor SL, Eller FJ, King JW (1997) A comparison of oil and fat content in oilseeds and ground beef- using supercritical fluid extraction and related analytical techniques. Food Res Int 30(5):365–370
Tezel A, Hortacsu A, Hortacsu O (2000) Multi-component models for seed and essential oil extraction. J Supercrit Fluids 19:3–17
Tundo P, Anastas P (2000) Synthetic pathways and processes in green chemistry. Introductory overview. Pure Appl Chem 72:1207–1228
Turner C, King JW, Mathiasson L (2001) Supercritical fluid extraction and chromatography for fat-soluble vitamin analysis. J Chromatogr A 936:215–237
Turner C, Eskilsson CS, Bjorklund E (2002) Collection in analytical-scale supercritical fluid extraction. J Chromatogr A 947:1–22
U.S.EPA Method 3561 Supercritical fluid extraction of polycyclic aromatic hydrocarbons. http://www.epa.gov/SW-846/pdfs/3561
Veriansyah B, Park TJ et al (2005) Supercritical water oxidation of wastewater from LCD manufacturing process: Kinetic and formation of chromium oxide nanoparticles. J Supercritl Fluids 34(1):51–61
Wang SM, Ling YC, Giang YS (2003) Forensic applications of supercritical fluid extraction and chromatography. Forensic Sci J 2:5–18
Wardencki W, Namieśnik J (2002) Some remarks on gas chromatographic challenges in the context of green analytical chemistry. Pol J Environ Stud 11:185–187

White C, Burnett J (2005) Integration of supercritical fluid chromatography into drug discovery as a routine support tool: II. Investigation and evaluation of supercritical fluid chromatography for achiral batch purification. J Chromatogr A 1074:175–180

White A, Burns D, Christensen TW (2006) Effective terminal sterilization using supercritical carbon dioxide. J Biotechnol 123(4):504–515

Yamaguchi T (1998) Structure of subcritical and supercritical hydrogen-bonded liquids and solutions. J Mol Liq 78(1–2):43–50

Yamini Y, Asghari-Khiavi M, Bahramifar N (2002) Effects of different parameters on supercritical fluid extraction of steroid drugs, from spiked matrices and tablets. Talanta 58:1003–1010

Yang Y, Strickland Z et al (2011) Industrial application of green chromatography-I. Separation and analysis of niacinamide in skincare creams using pure water as the mobile phase. Talanta 84(1):169–174

Yeo SD, Kiran E (2005) Formation of polymer particles with supercritical fluids: a review. J Supercrit Fluids 34(3):287–308

Zúñiga MC, Jover E et al (2009) Development of a methodology for the simultaneous determination of inorganic and organolead compounds using supercritical fluid extraction followed by gas chromatography-mass spectrometry and its application to environmental matrices. Talanta 80(2):504–510

Chapter 4
High Performance Thin-Layer Chromatography

Dušanka Milojković-Opsenica and Filip Andrić

Abstract The current chapter emphasizes the use of high-performance thin-layer chromatography (HPTLC) as a green analytical separation alternative. The chapter provides historical development of thin-layer chromatography towards becoming modern, automated, high resolution technique in the form of high-performance thin-layer chromatography, and their further advances in miniaturization of chromatographic beds in the form of ultra-performance thin-layer chromatography (UPTLC). Also, following the three R philosophy (reduce, replace, reuse), and twelve principles of green chemistry, the chapter provides the short review of green sample pretreatment techniques coupled with the HPTLC. In addition, the use of green separation modalities such as reversed-phase chromatography, hydrophilic interaction chromatography and salting out thin-layer chromatography is emphasized. The use of thin-layer chromatography as a simple screening technique and a powerful sample pretreatment method is highlighted.

4.1 Introduction

Since the introduction of twelve basic principles of green chemistry by Paul Anastas (Anastas and Warner 1998; Anastas 1999), different areas of chemistry embraced the green philosophy. Although paradoxical, analytical chemistry that should be considered as the most clean of all chemistry disciplines somehow appeared not to be. Consumption of the toxic solvents and reagents used for sample treatment and analysis put more burdens on environment and in some cases these chemicals have been found more toxic than the analytes themselves. Therefore, the need for developing green analytical methodologies felt a necessity.

It can be assumed that the fruitful seeds for development of green analytical chemistry (GAC) have been planted by Mallisa who presented his ideas about changing paradigms in analytical chemistry in Paris in 1987 during seventh Euroanalysis. Although the analytical community has been environmentally sensitive for a long time and the idea of improving methods by reducing the consumption of

D. Milojković-Opsenica (✉) · F. Andrić
Faculty of Chemistry, University of Belgrade, Studentski trg 12-16,
11000 Belgrade, Serbia
e-mail: dusankam@chem.bg.ac.rs

Inamuddin, A. Mohammad (eds.), *Green Chromatographic Techniques*,
DOI 10.1007/978-94-007-7735-4_4, © Springer Science+Business Media Dordrecht 2014

solvents predates the theoretical developments, the first papers relating GAC methods appeared in 1995 (De la Guardia and Ruzicka 1995). The three R philosophies (Replace, Reduce and Reuse) have been embraced in developing analytical procedures, and mostly applied to the sample pre-treatment procedures, as well as signal production and acquisition. The extraction procedures that use minimal amounts of solvents, or those that replace the toxic with the more environmentally benign or the non-toxic ones, such as: solid phase extraction (SPE), solid-phase microextraction (SPME), accelerated solvent extraction (ACE), liquid-phase micro-extraction (LPME), microwave-assisted extraction (MAE), or those that involve the use of ultrasound and supercritical fluids, have been developed and implemented in majority of GAC procedures. Before the development of micro-, nano-flow and ultra-performance liquid chromatography (UPLC) and capillary HPLC, at the turn of the century, even with the green sample pre-treatment methodologies, separation techniques such as liquid chromatography consumed significant amounts of toxic organic solvents. Reduction of the sorbent particle size resulted in the development of UPLC technique that uses shorter columns, higher pressure and provides significant shortening of the time of analysis, and reducing the amount of waste produced. The main motto was: fast chromatography, green chromatography. Development of capillary and nano flow HPLC significantly reduced the mobile phase consumption and make chromatographic process even more environmentally friendly. In the meantime HPLC techniques that use, so-called reversed-phase modality that employs mobile phases based on water and methanol or ethanol as organic modifier, replacing toxic volatile organic solvents, such as acetonitrile, gained more popularity (Ribeiro et al. 2004; Capello et al. 2007; Welch et al. 2009). Superheated water chromatography (SWLC) may be considered as one of the greenest examples of modern RPLC application (Greibrokk and Andersen 2003; Guillarme et al. 2004; Smith 2006), but it has some drawbacks because high temperature water (> 100 °C) is highly reactive and capable to hydrolyze classical octadecyl- and octyl- modified silica. Therefore, some temperature resistant packing materials such as polystyrene-divinylbenzene or zirconia-based materials are needed (Hartonen and Riekkola 2008).

The ACS Green Chemistry Institute has developed "greenness" criteria for environmental methods as a way to identify analytical chemistry methods that use minimal harmful solvents or safer chemicals, and produce little waste. These criteria have been applied to the National Environmental Methods Index (NEMI) which is a free Internet-searchable database of environmental methods located at www.nemi.gov.

Since its beginnings, thin-layer chromatography required small amounts of samples and solvents, with little or no sample preparation. Despite the advances in the automation, miniaturization and standardization of the overall chromatographic process, in the context of contemporary green analytical procedures modern TLC still does not take the appropriate place. Surprisingly it is strongly neglected to such extent that there is no single HPTLC procedure qualified in the NEMI data base.

4.2 High Performance Thin-Layer Chromatography

Among widely used chromatographic methods, thin-layer chromatography (TLC) is the simplest to perform (Sherma 2003). This rapid, sensitive, and economical analytical method can be used for qualitative and quantitative analysis of various organic and inorganic substances. TLC is a liquid chromatographic technique which differs from high performance liquid chromatography (HPLC) in a configuration of stationary phase while the separation mechanisms for these two methods are the same. Because of that, TLC can be used to predict retention behavior of different substances in the conditions of liquid column chromatography. Apart from an analytical chromatographic method, TLC can also be utilized as a preparative technique.

Analogous to other chromatographic systems, the TLC system is consisted of three principal and inseparable elements: the stationary phase—a sorbent, the mobile phase—a developing solvent, and the separated substance (an analyte or solute). The TLC sorbents, i.e. solid, porous materials with relatively high specific surface area, are in the form of a thin layer situated on a plane surface (commonly glass, metal or plastic plate). The mobile phase is liquid, comprising one or more solvents.

Two opposite forces: the elution force of mobile phase and the attraction of stationary phase, i.e. the ability of the sorbent to retain separated substances on its own surface active centers, determine the movement of substances under TLC conditions. At a given temperature, chromatographic process is governed by intermolecular interactions such as the solute–stationary phase, solute–mobile phase, and mobile phase–stationary phase. These interactions include hydrogen bonding, electrostatic forces such as ion–ion and ion–dipol interactions, electron-pair donor–electron-pair acceptor (charge-transfer) interactions, Van der Waals forces which include dipole–dipole (Keesom) interactions, dipol–induced dipole (Debay) interactions, and instantaneous dipol–induced dipole (London) interactions (Sherma 2003; Miller 2005; Vailaya 2005).

Retention of an analyte in TLC is described by so-called retardation factor, R_F, defined as the ratio of single zone distance from the starting point to the center of chromatographic zone (x) and the mobile phase distance from the starting point to the solvent front (f):

$$R_F = \frac{x}{f} \tag{4.1}$$

The R_F takes values between 0 (solute remains on the starting point) and 1 (solute migrates with the solvent front) and it represents the principal quantity used for the identification of separated substances (Kowalska 2003). Another parameter, R_M, with physical meaning corresponding to the log$k$ in HPLC, and related to R_F value according to the Eq. 4.2 has been proposed by Bate-Smith and Westall (1950).

Table 4.1 Characteristics of conventional TLC and HPTLC

Parameter	Conventional TLC	HPTLC
Plate size (cm)	20 × 20	10 × 10
Layer thickness (μm)	250	100 or 200
Average particle size (μm)	10–15	5–7
Particle-size distribution (μm)	0–25	0–10
Average plate height (μm)	35–75	23–25
Number of samples	10	18–36
Sample volume (μL)	1–5	0.1–0.2
Solvent volume (mL)	50	5–10
Development distance (cm)	10–15	3–5
Development time (min)	30–200	3–20
Limit of detection (ng)		
Absorbance	1–5	−0.5
Fluorescence	0.05–0.1	0.005–0.010

$$R_M = \log\left(\frac{1}{R_F} - 1\right) \tag{4.2}$$

The intensive development of TLC as a method of analysis of mainly organic substances started in the middle of the 20th century. Namely, the TLC technique was introduced in 1938 by Izmailov and Shraiber as so-called "spot chromatography" (Siouffi 2005). An actual development of TLC started in 1950s with Egon Stahl who introduced the term "thin-layer chromatography", devised the fabrication of the first commercial TLC plates (produced by Merck and promoted in 1958), and standardized the method. In 1956 the first paper entitled "Thin layer chromatography" was published (Stahl 1956). The 1960s were marked by introduction of scanning densitometry in TLC practice, i.e. the development of quantitative applications by Kirchner (1967). TLC became widely-used low-cost chromatographic method for analysis of complex mixtures of various organic and inorganic substances. Further improvement in TLC was resulted with commercial production of precoated TLC plates with uniform, fine-particle layers in the 1970s. The use of silica gels with a small particle diameter as well as the development of different surface-modified sorbents around 1978 and the introduction of layers prepared from spherical particles around 1990, has been a driving force for intensive development of theory, practice, and instrumentation that led to the methods termed high-performance thin-layer chromatography (HPTLC) and instrumental HPTLC (Sherma 2003). Some important characteristics of conventional TLC and HPTLC are comparatively listed in Table 4.1 (Vailaya 2005; Poole 2003; Poole and Poole 2011).

HPTLC is an advanced form of instrumental TLC that means the use of high performance sorbent layers, adopted instrumentation, standardized methodology for development, optimization, documentation, as well as the use of validated methods (Shewiyo et al. 2012). The HPTLC technique is widely used in many areas of science and technology: the pharmaceutical and food industries, cosmetology, biochemistry, clinical chemistry, environmental sciences. Literature search reveals a large number of published papers that describe various applications of this chromatographic technique; in numerous of them the importance, challenges and opportunities of the HPTLC have been reviewed (Vailaya A 2005; Siouffi 2005; Kirchner 1967).

Modern thin-layer chromatography is an instrumental technique that is comparable by its accuracy and precision with both gas and high-performance liquid chromatography. Moreover HPTLC has several preferences over HPLC and other techniques (Vailaya 2005; Gupta et al. 2012; Morlock and Schwack 2008). Some of the advantages are:

- small amount of sample is needed for analysis,
- better precision and accuracy caused by simultaneous analysis of both samples and standards under the same conditions,
- low sensitivity to impurities,
- there is no possibility of interference from previous analysis as fresh stationary phase is used for each analysis,
- there are wide choices of adsorbents and developing solvents,
- mobile phase consumption per sample is extremely low,
- instrumentation is simple, relatively inexpensive and easy to handle,
- many samples can be separated in parallel on the same plate resulting in a high through-put, and a rapid low-cost analysis,
- the possibility of simultaneous separation with the use of various mobile phases (the HPTLC Vario developing chamber from Camag),
- the multidimensional capability,
- the easy and rapid optimization,
- a short time of analysis,
- the cost of quantitative analysis is low (about 35 % of the cost of HPLC analysis).

The continuous development of liquid chromatographic methods and, therefore HPTLC, resulted in application of monolithic materials in chromatography (Poole 2011; Frolova 2011). This revolutionary improvement led to the introduction of so-called ultra-performance thin-layer chromatography (UPTLC) at the beginning of 21st century (in 2001). UPTLC with either monolithic or nanostructured sorbents appears more favourable many advantages in comparison to HPTLC because of faster separations (1–6 min) over shorter distances (1–3 cm) due to its thinner stationary phases (10 μm) with finer pore sizes (1–2 μm macropores and 3–4 nm mesopores). Consequently, the consumption of mobile phase reduced to many fold. In most cases, UPTLC plates are capable to exhibit lower limit of detection. However, they provide lower resolution due to shorter development lengths and lower available specific surface area (Bezuidenhout et al. 2008; Clark and Olesik 2009; Jim et al. 2011). Therefore, UPTLC is greener than HPTLC, but the later is still the most widely used planar chromatographic method.

4.3 Sample Preparation in HPTLC

The best GAC method is the one that does not require any sample preparation. In that way HPTLC has been applied to the variety of samples with little or no demands on the tedious preparation step. Since the entire chromatographic process takes place on the plate, which can be preserved for future inspection and quantification, the analysts need to take a little care about all impurities and interferences in the sample. Carefully designed and optimized separation steps can move all impurities non-selectively towards the solvent front or completely keep them at the starting position, leaving the target analytes free from interference. Even if it is impossible to separate interfering compounds from target substances, still choosing adequate detection modality, including the proper selection of derivatization reagents can significantly increase selectivity. For such reckless attitude towards putting the impurities along with sample target analytes, one particular feature of contemporary HPTLC is responsible, that is, each de novo chromatographic experiment requires the use of new chromatographic bed.

In cases where a sample pretreatment is necessary, the use of the simplest, quick and cheap option is obvious. For solid samples the extraction of active ingredients is mostly carried out with ultrasound assisted solvent extraction (UAE), microwave assisted solvent extraction (MAE), accelerated solvent extraction (ASE) and supercritical fluid extraction (SFE). For further extract purification, solid-phase extraction (SPE) is mostly used technique. Application of ultrasound makes the process of extraction faster and accompanied with higher yields of target analytes. It usually do not require less amounts of solvents, that required in classical Soxhlet extraction, but can be carried out at lower temperatures while the technique itself is less tedious and less time consuming.

The MAE has been applied to the extraction of organic compounds from different types of matrices. It employs less organic solvent and a shorter extraction time than traditional extraction methods, even when compared with UAE (Bhope et al. 2011). The main advantages of MAE are short extraction time, reduction in the amount of sample required, high sample throughput, and high level safety, since it does not require the use of hazardous chemicals and performed in closed reactors.

In the same way, ASE, also known as pressurized solvent extraction (PSE), pressurized fluid extraction (PFE), pressurized liquid extraction (PLE), and solvolytic extraction, is a solid–liquid extraction process performed at high temperatures (50–200 °C) and high pressures (10–15 MPa). ASE is a form of PSE similar to SFE, although, in ASE, the extraction is carried out under pressure to maintain the solvent in its liquid state at high temperature, but always below its critical condition. Although the solvent used in ASE is usually organic, pressurized hot water can also be used. Nowadays, ASE is considered a potentially, attractive, alternative technique for extracting organic compounds from environmental or biological matrices because its main advantages over traditional extraction methods being dramatic decreases in the amount of solvent used and the shortening of extraction time. (Sharma et al. 2007) compared several techniques (UAE, MAE, SFE and Soxhlet method) for extraction of vanillin and related phenolic compounds from dried and pulverized plant samples of *Vanilla planifolia*. They used reversed-phase high performance thin-layer chromatography (RP-HPTLC) as the useful way to quantify active ingredients and to assess the most effective sample preparation method. According to their findings, ASE and MAE give the best yield of target analytes and in the same time significantly shorten the analysis time and solvent consumption. This is a nice example of green analytical method, since the entire chromatographic step require small amount of methanol/ water/ isopropanol/ acetic acid mixture for simultaneous analysis of several sample extracts, while consuming only 20 ml of dehydrated ethanol for sample extraction.

The SFE offers an attractive alternative to overcome the unfavorable effect of non-polar organic solvents employed in extracting non-polar compounds. This method is useful because: (a) it can achieve high concentrations; and (b) it is quantitative, it is fast, it is simple, and, selective. SFE is also an environmentally-friendly analytical methodology that can be automated easily and completely. Mei-Chih Lin et al. (1999) used supercritical fluid extraction for isolation of flavonoids from *Scutellariae radix* plant. They varied the composition of extracting solvent by addition of polar methanol to relatively non-polar liquid carbon-dioxide in order to increase the overall polarity of extracting solvent. They also studied the impact of temperature and pressure applied, and found that optimal condition were reached under the pressure of 200 bar and temperature of 50 °C with the solvent mixture 70% methanol in water and liquid CO_2 in the ratio 3:20. They also compared the ultrasound assisted solvent extraction and other conventional techniques and found that compared to SFE, sonication can lead to higher yield of crude extract. In addition to GC/MS, they used conventional NP-HPTLC to separate baicalin, baicalein and wogonin as target analytes and to follow the optimization of extraction process using the mixture of *n*-butanol–water–glacial acetic acid (7:2:1, v/v/v).

For liquid samples the most useful extraction method is liquid-liquid microextraction (LLME). Essentially, LLME is miniaturized liquid–liquid extraction in which the analyte moves between the bulk aqueous phase and a very small volume of organic solvent. Recent modification in the methodology is the use of a single droplet of solvent, suspended at the tip of a needle and exposed to the sample solution. Faraji et al. (2011) used LLME for determination of phenolic compounds in water. Target analytes were extracted using microliter volume of 1-undecanol. Analytes were further separated on RP-18 modified HPTLC silica gel plates using a mixture of methanol—water (2:1).

Solid-phase extraction as a cleanup procedure that usually follows a rough sample pretreatment methods and crude isolation of target compounds is used in order to get rid of major interferences. Modified silica (C-18) and other hydrophobic materials have been mostly used as sorbents, along with some amphiphilic materials. These are usually applied to plant—herbal analysis, environmental samples, drugs and pharmaceutical formulations (Pobłocka-Olech and Krauze-Baranowska 2008; Bazylak 2000). In that way (Tuzimski 2010) used commercially available RP18/SDB-1 SPE cartridges for extraction of pesticides in water that was afterwards followed with HPTLC separation step coupled with DAD detection. Similarly Poblochka-Olech and Krauze-Baranowska (Pobłocka-Olech and Krauze-Baranowska 2008) used the commercially available LichroprepTM RP-18 SPE cartridges for herbal extract purification in for further HPTLC analysis of procyanidins.

Instead of using cartridges Oellig and Swack (2011) developed automated planar chromatographic system for powerful sample clean-up, called high-throughput planar solid-phase extraction (HTpSPE). Actually, they brought the old ideas incorporated in the preparative thin-layer methodology to its renascence. They used thin-layer chromatography to completely separate pesticides from matrix compounds and to focus them into a sharp zone, which were then extracted by the TLC–MS interface. The extracts were nearly free of interference and free of matrix effects, as shown for seven chemically representative pesticides in four different matrices (apples, cucumbers, red grapes, and tomatoes).

Ultrasound assisted extraction has been often used in combination with HPTLC separation because it improves the extraction process, shortens the extraction time and reduces the amount of solvents. Rezić and co-workers (Rezić et al. 2005) used ultrasonic extraction to isolate and determine the traces of pesticides, atrazine and simazine from honey samples. They used various solvents such as: acetone, diethyl-ether, chloroform, hexane, benzene, acetonitrile and dichloromethane in order to optimize extraction step. After sonication, and filtration, organic phase was separated and evaporated on rotary evaporator, to dryness. The residue was reconstituted in methanol. They combined the Quantitative HPTLC analysis while searching for optimal extraction conditions.

Wu and coworkers (Wu 2001) evaluated ultrasonic extraction of ginseng saponins from ginseng roots and cultured ginseng cells in combination with HPTLC. They used probes for direct and indirect sonication. 200 mg of ginseng sample was mixed with 15 ml of extraction solvent. After careful temperature control and total sonication power delivery, the extract was centrifuged, supernatant evaporated and residue reconstituted in 0.5 ml of acetonitrile and applied to TLC plate (silica gel 60 F_{254} 0.25 mm); the plates were developed with butanol—ethyl-acetate—water (4:1:5) mixture.

Panadda Phattanawasin and co-workers (Phattanawasin 2012) developed a simple HPTLC image analysis method for determination of sibutramine in adulterated slimming herbal formulations. They used a simple sonication procedure with 20 ml of methanol for each single herbal capsule dosage. The resultant solution was directly applied on silica gel 60 F_{254} HPTLC plate, that was developed with mixture of toluene–n-hexane–diethylamine (9:1:0.3, v/v/v). Detection of target compound was done after exposing the plate to Dragendorff reagent. Image of chromatogram was used for quantitatification purposes (Table 4.2).

Table 4.2 The list of methods for sample pretreatment commonly used in combination with HPTLC

Sample pretreatment	Type of sample	Short method description	HPTLC mode	HPTLC Stationary and/or mobile phase	References
Solid phase extraction (SPE)	Methanol extract of dried and pulverized plant material	Sorbent LichroprepTM RP-18 cartridges eluted with methanol–water 7:93 v/v	NP-HPTLC	5 cm × 10 cm silica gel 60 eluted with chloroform–ethanol–formic acid (50:40:6 v/v/v)	Pobłocka-Olech and Krauze-Baranowska (2008)
Solid phase extraction (SPE)	Solution of drug formulation -3,4-ethylene dioxymethamphetamine	C18 SPE Eluted with 2 ml water. Analytes eluted with ten portions of 50 µl methanol	NP-HPTLC	Various solvent mixtures methanol (9: 1 v/v), acetonitrile: chloroform (1: 1 v/v)	Kochana et al. (2006)
Solid phase extraction (SPE)	Pesticides in water samples	C18/SDB-1 [C18 500 mg on top + styrene divinylbenzene copolymer 200 mg on ottom/6 mL] eluted with MeOH–H2O (5 + 95, v/v) and dried and eluted again with pure MeOH	NP-HPTLC	HPTLC silica gel eluted with ethyl acetate–n-heptane (40 + 60, v/v)	Tuzimski (2010)
Solid phase extraction (SPE)	Norethindrone acetate and dydrogesterone commercially available tablets	C18 cartridge Eluent 10 ml of pure MeOH with flow rate of 0.1 ml/min	NP-TLC	In-house prepared Silica gel and impregnated with Cu^{2+}, Fe^{2+}, Co^{2+}, Zn^{2+}, and Ni^{2+} salts Mobile phase: *n*-hexane—*n*-butanol (90:10, v/v)	Ali et al. (2012)
Ultrasound solvent extraction (USE)	Atrazine and simazine in honey samples	Sonication with benzene:water = 1:1 (v/v) mixture	NP-HPTLC	HPTLC Silica Gel 60 F_{254} Mobile phase hexane–chloroform–acetone (60:25:15, v/v/v)	Rezić et al. (2005)
Ultrasound solvent extraction (USE)	Lipids in marine mucilage samples	Sonication of 300 mg sample with diethyl ether for 1 min	NP-TLC	Silica gel Mobile phase hexane -diethyl ether -acetic acid (97.8:2.0:0.2, v/ v)	Mecozzi et al. (2002)
Ultrasound solvent extraction (USE)	Dried and pulverized plant material—rotenoids boeravinone B and E	Sonication of 1 g sample sonicated with 250 mL ethanol–water (50 + 50, v/v) for 60 min	NP-HPTLC	Silica gel 60 F_{254}, aluminum precoated Mobile phase: toluene–ethyl acetate–acetonitrile–formic acid (15 + 3 + 1 + 0.75, v/v/v/v)	Bhope et al. (2011)

Table 4.2 (continued)

Sample pretreatment	Type of sample	Short method description	HPTLC mode	HPTLC Stationary and/or mobile phase	References
Microwave assisted extraction	Dried and pulverized plant material—rotenoids boeravinone B and boeravinone E	1 g sample extracted with 250 mL ethanol–water (50+50, v/v) for 10 min	NP-HPTLC	Silica gel 60 F_{254}, aluminum precoated Mobile phase: toluene–ethyl acetate–acetonitrile–formic acid (15+3+1+0.75, v/v/v/v)	Bhope et al. (2011)
Microwave assisted extraction	Powdered vanilla pods. Determination of vanillin and related phenolic compounds	Extracted at 60 °C during 20 min with dehydrated ethanol as extraction solvent	RP-HPTLC	HPTLC, RP-18 modified silica F_{254S} plates methanol/water/isopropanol/acetic acid (30:65:2:3, by volume)	Sharma et al. (2007)
Pressurized fluid extraction (PFE) or Accelerated solvent extraction (ASE)	Powdered vanilla pods. Determination of vanillin and related phenolic compounds	A weighed amount (5 g) of powdered vanilla pads extracted with dehydrated EtOH, at 10.342 MPa pressure, and 60 °C temperature	RP-HPTLC	HPTLC, RP-18 modified silica F_{254S} plates methanol/water/isopropanol/acetic acid (30:65:2:3, by volume)	Sharma et al. (2007)
Liquid-liquid microextraction (LLME)	Phenolic compounds in water samples	Microliter volume of 1-undecanol floating on the surface of water sample	RP-HPTLC	C-18 modified silica commercially available HPTLC plate Mobile phase: mixture of methanol–water 2:1 (%, v/v)	Faraji et al. (2011)
Supercritical fluid extraction (SFE)	Pulverazied plant material *Scutellariae Radix* Determination of flavonoids	Methanol or 70 % methanol and 20 ml of liquid carbon dioxide was used as extraction solvent	NP-HPTLC	Silica gel Merck 60 F 254 Mobile phase butanol–water–glacial acetic acid (7:2:1, v/v/v) tetrachloromethane–acetone (5:3, v/v)	Lin et al. (1999)

4.4 Green Separation Modalities in HPTLC

Although, the modern HPTLC seems to be a greener technique than related separation methods such as HPLC, still it can be discussed in the terms of three R philosophy (Replace, Reduce, Reuse), to make the existing methods more environmental friendly. In the realm of liquid column chromatography, it can be achieved either through replacement of the toxic solvents such as benzene, toluene, acetonitrile or tetrahydrofuran, with less toxic ones such as ethanol, or the non-toxic alternatives such as a superheated water. The other way is to reduce the overall consumption of solvents and waste production during chromatographic separation step by shortening the analysis time or reducing the mobile phase flow. In order to shorten the time of analysis, the production of shorter columns with better efficacy is required. Finally, a novel chromatographic technique so-called ultra-performance liquid chromatography emerged, with particles size less than 2 μm and the column dimensions of 100×2.1 mm. Total consumption of solvents per analysis in this particular case was five folds smaller than in the case of conventional HPLC. The other alternative was a reduction in mobile phase flow while maintaining the separation efficacy at the same level. Finally, this idea has evolved in development of HPLC micro-flow, capillary flow and nano-flow systems. It can be stated that similar concept was considered in advancing of TLC methodologies, through development of new sorbents with uniform and narrower particle size distribution, chemically bonded stationary phases that provide tremendous separation versatility and automation of the entire chromatographic process. HPTLC provides fair separation and better quantitation compared to modern HPLC systems by consuming significantly smaller amounts of solvents in the same time. Still is it the smaller amount of consumed solvents enough? Can it be done more to make HPTLC even greener technique?

4.4.1 "Three R" Philosophy—Replacement of Toxic Solvents with Environmental Friendly Mobile Phases

Most of the work reported of HPTLC separations has been performed by normal-phase mode (NP), that is, the polarity of the stationary phase, usually silica gel, is greater than the polarity of mobile phase (commonly used the mixtures of volatile and toxic organic solvents, such as: hexane, benzene, diethyl ether, chloroform, dioxane, formic acid).

By embracing greener separation modalities and replacing toxic solvents with less toxic ones, several separation modalities that use water mixtures with some of organic solvents such as methanol, ethanol, acetonitrile, or pure water with addition of salts become more popular. Among these techniques we would like to emphasize: the reversed-phase chromatography (RP), chromatography of hydrophilic interactions (HILIC) and salting-out chromatography. These techniques became popular through ever-expanding choice of sorbents and their unique selectivities became available when modifications began to be made on silica gel by way of siloxane bonding. The main advantages of these modifications are: (a) Phase stability—there is no bleeding of the thin layer of stationary phase during chromatographic process which was the case with coated layers; (b) the possibility to apply broad range of retention mechanisms to the separation process.

4.4.1.1 Reversed-Phase Chromatography

Many chromatographic separation problems can be solved by using hydrophobic interactions of a stationary phase with compounds of appropriate molecular structure. Sorbents that are suitable for this task are called reversed-phase (RP) materials. Reversed phase means that the relative polarities of the stationary and mobile phase are reversed compared with the situation in adsorption chromatography described as normal-phase chromatography. Hydrophobic character of stationary phase can be modified in two ways: i) by choosing proper alkyl group to bind the siloxane ending or ii) by degree of modification.

The most common matrix for obtaining hydrophobically modified sorbents is porous silica. Silica gel with 6 nm pores is usually used in HPTLC. The popular functional groups are ethyl (RP-2), octyl (RP-8), dodecyl (RP-12), octadecyl (RP-18) and phenyl ligand where hydrophobic character increases from RP-2 to RP-18. The steric hindrance increases with the increase in the alkyl chain length thereby influencing the density of active centers. Hydrophobicity plays an important role in thin-layer chromatography and in the same time makes it greener alternative because mainly aqueous mobile phases are used. Therefore, the range of organic modifiers is limited to those that are miscible with water. The increase in the water content creates few problems. If the repulsive—hydrophobic forces are stronger than capillary forces, the mobile phase does not wet the stationary phase layer. In order to overcome this problem only one portion of silanol groups are modified, while a certain degree of these groups remains free. This resulted in stationary phase that still exhibit RP properties while in the same time may be used with pure water as an eluent. The most commonly used mobile phases are mixtures of organic modifiers such as methanol, acetonitrile, tetrahydrofuran, acetone and dioxane, with the emphasis that these more toxic solvents may be replaced with ethanol as greener component in spite of one fact that it has some disadvantages over acetonitrile. The amount of solvent used per development has been about 2–5 ml, if horizontal technique is used, which results in production of minimal amounts of waste. In this light, Miszczyk and Pyka (2006) have compared the NP and RP conditions for separation of fifteen urea derived herbicides. They finally concluded that some groups of studied compounds cannot be satisfactorily separated in NP mode, while reversed-phase modality provides good separation for all studied groups. The use of methanol—water mixtures makes this separation mode preferable over the classical NP step that consumes toxic benzene-methanol mixtures.

The nice example of application of RP-HPTLC in testing of environmental samples is from Kutsch and Schoen (2000) who used commercially available HPTLC plates without fluorescence indicator for screening of soil samples for total amount of polyaromatic hydrocarbons. They found that it is more environmental friendly to use minimal amounts (20 ml) of ethanol-dichloromethane (1:1) mixture as a mobile phase, for screening purposes, while the use of ACN gave better separations. It reflects that HPTLC technique is most suitable for screening purposes since the screening of samples significantly reduces the total number of sample analyzed by standardized methods—and therefore significantly reduces the overall environmental burden, as well as the total cost.

Krzek et al. (2005) used simple RP-18 HPTLC method for determination of enantiomeric forms of ibuprofen, by adding chiral modifier (1 % water solution of ß-cyclodextrin) in mobile phase (ß-CD sol.—methanol, 15:1). They achieved good accuracy (99.4 %) as well as low limit of detection (1 μg) (Table 4.3).

Beside its application in analytical separation and determination of drugs, herbal active compounds, food analysis, and environmental samples, reversed-phase thin-layer chromatography has a long history in lipophilicity assessment, as well as QSRR studies of biologically active compounds (Wang and Zhang 1999; Giaginis and Tsantili-Kakoulidou 2008; Tsantili-Kakoulidou 2005). The most commonly used mobile phases are the mixtures of some organic modifier such as methanol, acetonitrile, acetone, or dioxane with water. The basic idea that lies behind the entire concept is the linear relationship between retention parameters (expressed as R_M values) of studied compounds and mobile phase composition, i.e. the percentage of organic modifier. The linear relationship results in three features that are proportional to the octanol-water partition coefficient (log K_{OW}), the intercept (R_M^0), slope (b) and the C^0 that is the ratio of the last two ($-R_M^0/b$). All three parameters have their physical explanation. The intercept should be interpreted as the retention of a compound at the zero content of organic modifier in the mobile phase. Slope is proportional to the specific hydrophobic area of the molecule accessible to the alkyl chains of stationary phase. Sarbu et al. (2002) showed that the linear combination of retention factors in several chromatographic systems is better correlated with experimentally determined lipophilicity parameters than any of the aforementioned parameters.

Also typical reversed-phase thin-layer chromatographic systems were used in combination with various calculation procedures to assess the soil-water partition coefficient normalized to the organic carbon (Log K_{OC}) (Andrić et al. 2010). Compared with official (OECD) HPLC procedure (OECD guideline for the testing of chemicals, No 121) the proposed method consumes much lower amounts of solvents (4–5 ml per analysis) (Table 4.4).

4.4.1.2 Hydrophilic Interaction Chromatography (HILIC) in HPTLC

Hydrophilic interaction chromatography has been introduced by (Alpert 1990). It represents chromatographic separation that uses polar mobile phases, usually mixtures of organic solvents with water (similarity with typical RP mode) and polar stationary phases such as silica (similarity with typical NP mode). HILIC is suitable for compounds that are extremely polar (sugars, amino acids, peptides, inorganic

Table 4.3 Examples of application of reversed-phase separation modality as a green method in HPTLC

Analysis	Sample	HPTLC mode	Stationary phase	Mobile phase	Reference
Drug analysis	Levodopa, carbidopa, and entacapone in combined tablet dosage form	RP	RP-18 modified silica 60	Acetonitrile/n-butanol/water/triethylamine (0.5:9.5:1:0.001, v/v/v/v)	Gandhi and Mehta (2011)
Environmental sample analysis	Urea pesticides	NP	NP: Silica gel 60 F_{254}	NP: benzene/methanol and benzene/ethanol (85–100 % of less polar component)	Miszczyk and Pyka (2006)
		RP	RP: RP-18WF_{254} coated on glass	RP: methanol/water (40–90 % of methanol in steps of 5 %) and acetonitrile/methanol (1:1, v/v)	
Analysis of herbal extracts	Apigenin, luteolin, quercetin, quercitrin and rutin	RP-HPTLC	RP-18 F_{254} HPTLC plates	Double run with: a) 5 % formic acid in water and methanol (70:30) b) 5 % formic acid in water and methanol (50:50)	Bhandari et al. (2007)
Drug analysis	S(+) and R(−) ibuprofen	RP-TLC	RP-18 modified silica gel F_{254} coated on aluminum sheets	1 % water solution of β-cyclodextrin and methanol (15:1, v/v)	Krzek et al. (2005)
Environmental soil samples (screening)	Polyaromatic hydrocarbons (PAH)	RP-HPTLC	RP18 HPTLC without fluorescence indicator	ethanol-dichloromethane (1:1, v/v)	Kutsch and Schoen (2000)

Table 4.4 The short list of the recently published papers dealing with HPTLC application in QSRR and lipophilicity assessment

Application	Compounds	Type of TLC system used	Stationary phase	Mobile phase	References
Lipophilicity determination	Pesticides	RP-HPTLC	RP-18, RP-8 and CN stationary phases	methanol-water in various proportions	Nascu-Briciu and Sarbu (2010)
Lipophilicity determination	Derivatives of fourteen 1,3-benzoxazol-2(3H)-ones	RP-HPTLC	Precoated C18 F254 plates	Mixts. of methanol-water and aminoacetic acid buffer, pH 2.67 and 11.6	Skibinski et al. (2011)
Lipophilicity determination	1-[3-(Y-alkoxyphenylcarbamoyloxy)-2-hydroxypropyl]-4-(2-methylphenyl)piperazine chlorides	RP-HPTLC	–	–	Malik et al. (2011)
QSRR	Arylpiperazine derivatives	RP-HPTLC	HPTLC RP18 Silica F254	Various mixtures of methanol-water, dioxane-water and dimethylsulfoxide-water.	Trifkovic et al. (2010)
QSRR	Cholic acid derived cis-trans isomeric bis-steroidal tetraoxanes	RP-HPTLC	HPTLC RP18 Silica F254	Various mixtures of methanol-water, dioxane-water and acetone-water	Šegan et al. (2011)
QSRR	N-substituted 2-alkylidene-4-oxothiazolidines	RP-HPTLC	HPTLC RP18 and CN modified silica	–	Dabić et al. (2011)
Lipophilicity determination	Fourteen s-triazine derivatives	RP-TLC	Impregnated silica gel	Different mixtures of water-acetone, water-acetonitrile, water-dioxane, water- tetrahydrofuran, water-methanol and water-ethanol	Jevric et al. (2010)

complex compounds etc.), that are completely unretained in classical reversed-phase systems, while stay strongly retained under typical normal-phase conditions.

Several materials may be used as stationary phase in HILIC such as: bare silica, or polar chemically bonded as cyanopropyl, aminopropyl, diol or different ion-exchange and zwitterionic phases (Jandera 2008). Stationary phase is generally hydrophilic but, depending on the composition of the mobile phase and pH of the chromatographic system, it may stay uncharged or change to positively or negatively charged particle surface. Therefore, different separation mechanisms may take place, such as: partition, adsorption or ion exchange (Jandera 2008).

Contrary to vast number of HILIC application in HPLC, only few examples could be found related to the high/performance thin-layer chromatography. Búriová et al. (2005) used amino modified silica gel plates and mobile phase MeCN: 0.025 % (4 mM) of ammonium formate solution (80: 20 v/v) to follow auto-radiolysis degradation products of 2-deoxy-2-[18F] fluoro-D-glucose. Mohammad and Laeeq (2011) developed an HILIC-HPTLC method for the determination of hexoses, pentoses and disaccharides in blood serum and pharmaceutical formulations such as cough and multivitamin syrups. They used bare silica gel 60 HPTLC plates as stationary phase and sodium deoxycholate (NaDC) solution as additive in acetonitrile (ACN) in the ratio (1:5 v/v) as a mobile phase. They achieved detection of sugars in nanogram level. Cieśla et al. (2011) used 2D TLC HILIC system to successfully resolve highly polar iridoids and triterpene saponins of the different Verbascum spp. flower extracts. They performed TLC separations on silica gel plates, with two different mobile phases used in the perpendicular directions. First, the analyzed samples were developed with AcOEt–MeOH–H_2O–25 % aqueous NH_3 (55:35:9:1, v/v/v/v) and redeveloped with MeOH–AcOEt–H_2O–HCOOH (10:90:26:22, v/v/v/v) in a perpendicular direction. The resolved compounds were visualized using the vanillin–sulfuric acid reagent. Milojković-Opsenica and coworkers (Radoičić et al. 2009; Shweshein et al. 2012) used different hydrophilic sorbents in order to characterize chromatographic behavior of Co(III)-complexes and their cis-trans isomers under various HILIC conditions.

4.4.1.3 Salting-Out Chromatography in HPTLC

Salting-out thin layer chromatography (SOTLC) could be regarded in a broader sense as reversed-phase chromatographic technique that uses the concentrated aqueous solutions of various inorganic salts as mobile phases and polar stationary phases such as bare silica or its chemical modifications, cellulose, polyamide etc. So far, solutions of ammonium sulphate have been most extensively used as an eluent in SOTLC, due to its high solubility in water and high salting out effect. Several separation mechanisms are proposed (Grover and Ryall 2005) but basically salts present in the mobile phase in high enough concentrations lower the solubility of even un-ionized organic compounds and force them to interact stronger with sorbents, therefore exhibiting stronger retention. Flieger et al. (2007) confirmed by the means of QSRR that the major contributing factor in separation mechanism is so-called hydrophobic interactions. They studied chromatographic behavior of eleven sulfonamides on bare silica TLC HPTLC plates using various salts ($(NH_4)_2SO_4$, $CaCl_2$, NaCl, etc.) at different concentrations. SOTLC found application in different studies of various drugs. In that way Aleksić et al., gave an insight into chromatographic behavior of some ACE inhibitors under SOTLC conditions (Aleksić et al. 2001). Also the same author studied a behavior of five myorelaxant drugs under SOTLC conditions using cellulose and alumina as stationary phase and aqueous solutions of ammonium-sulphate of different concentrations as mobile phase. It was established that hR_F values always decreased in parallel to increasing salt concentration. When cellulose was used as adsorbent, a linear relationship was observed between the R_M values and the ammonium sulfate content of the mobile phase. Regression data of the plots obtained were used to determine the lipophilicity parameters R_M^0 and C^0. Lipophilicity determined in this way was correlated with calculated log P values (Aleksić et al. 2003). Odović et al. (2005), studied behavior of five ACE inhibitors and their active degradation products in salting-out thin-layer chromatography on silica gel, cellulose, and polyacrylonitrile (PAN) with aqueous ammonium sulfate solutions of different concentration as mobile phases. They established linear relationship between retention and salt concentration, and further successfully used this relationship to predict lipophilicity of studied compounds. As a part of their long term studies of chromatographic behavior of different inorganic complex compounds Milojković-Opsenica and coworkers (Milojković-Opsenica et al. 1999) studied the effect of chelate ring size in diamino Co(III)-complex compounds. Živković-Radovanović and Vučković (2008) studied aqueous solutions of 28 different salts as a potential mobile phases for salting-out thin-layer chromatography, on silica gel, of a series of four mixed bis-aminocarboxylato-Cobalt(III) complexes.

4.5 Conclusion

Since of its beginnings, thin-layer chromatography required small amounts of samples and solvents, with little or no sample pretreatment. Compared to other analytical procedures thin-layer chromatography has been followed with significantly low costs and should be considered among the greenest analytical techniques. Despite the advances in the automation, miniaturization and standardization of the overall chromatographic process, in the context of contemporary green analytical procedures modern TLC still does not take the appropriate place. HPTLC combined with simple sample preparation methods that pace little burden on environment, such as solid phase extraction, liquid-liquid microextraction, ultrasound assisted extraction, etc. can provide quick, reliable, low cost, and above all green analytical methodologies. Sometimes, HPTLC can be applied as a powerful sample cleanup procedure, or used as a fast and robust screening method that further significantly reduces the number of runs of costly and environmentally demanding analytical procedures. Even in the separation step in the scope of HPTLC process something more could be done, by simple embracing separation modalities that enforce the use of more environmentally friendly mobile phases which is in the case of typical RP-HPTLC, HILIC or salting-out chromatography.

Acknowledgment The authors are grateful to the Ministry of Education, Science and Technological Development of the Republic of Serbia (Grant No. 172017) for financial support and to Mr. Petar Ristivojević for kindly technical help.

References

Aleksić MB, Agbaba DG, Milojković-Opsenica DM et al (2001) Salting-out thin-layer chromatography of some angiotensin converting enzyme inhibitors on different sorbents. Chromatographia 53(S):442–444

Aleksić M, Odović J, Milojković-Opsenica DM et al (2003) Salting-out thin-layer chromatography of several myorelaxants. J Planar Chromatogr—Mod TLC 16(2):144–146

Ali I, Hussain I, Saleem K et al (2012) Development of efficient SPE–TLC method and evaluation of biological interactions of contraceptives with progesterone receptors. Arabian J Chem 5:235–240

Alpert AJ (1990) Hydrophilic-interaction chromatography for the separation of peptides, nucleic acids and other polar compounds. J Chromatogr A 499:177–196

Anastas PT (1999) Green chemistry and the role of analytical methodology development. Crit Rev Anal Chem 29(3):167–175

Anastas PT, Warner JC (1998) Green chemistry: theory and practice. Oxford University Press, New York

Andrić FLj, Trifkovic JĐ, Radoičić AD et al (2010) Determination of the soil–water partition coefficients (log KOC) of some mono- and poly-substituted phenols by reversed-phase thin-layer chromatography. Chemosphere 81(3):299–305

Bate-Smith EC, Westall RG (1950) Chromatographic behavior and chemical structure. 1. Some naturally occurring phenolic substances. Biochim Biophys Acta 4:427–440

Bazylak G, Brózik H, Sabanty W (2000) Combined SPE and HPTLC as a screening assay of urinary cotinine from male adolescents exposed to environmental tobacco smoke. Pol J Environ Stud 9(2):113–123

Bezuidenhout LW, Brett MJ (2008) Ultrathin layer chromatography on nanostructured thin films. J Chromatogr A 1183:179–185

Bhandari P, Kumar N, Gupta AP et al (2007) A rapid RP-HPTLC densitometry method for simultaneous determination of major flavonoids in important medicinal plants. J Sep Sci 30:2092–2096

Bhope SG, Ghosh VK, Kuber VV (2011) Rapid microwave-assisted extraction and HPTLC photodensitometric method for the quality assessment of Boerhaavia diffusa L. J AOAC Int 94(3):795–802

Búriová E, Macásek F, Melichar F, Kropácek M, Procházka L (2005) Autoradiolysis of the 2-deoxy-2-[18F] fluoro-D-glucose radiopharmaceutical. J Radioanal Nucl Chem 264 (3):595–602

Capello C, Fischer U, Hungerbühler K (2007) What is a green solvent? A comprehensive framework for the environmental assessment of solvents. Green Chem 9:927–934

Cieśla Ł, Hajnos M, Waksmundzka-Hajnos M (2011) Application of hydrophilic interaction TLC systems for separation of highly polar glycosidic compounds from the flowers of selected verbascum species. J Planar Chromatogr—Mod TLC 24(4):295–300

Clark JE, Olesik SV (2009) Technique for ultrathin layer chromatography using an electrospun, nanofibrous stationary phase. Anal Chem 81:4121–4129

Dabić D, Natić M, Džambaski Z et al (2011) Quantitative structure-retention relationship of new N-substituted 2-alkylidene-4-oxothiazolidines. J Sep Sci 34(18):2397–2404

Faraji H, Saber-Tehrani M, Mirzaie A et al (2011) Application of liquid–liquid microextraction–high-performance thin-layer chromatography for preconcentration and determination of phenolic compounds in aqueous samples. J Planar Chromatogr 24(3):214–217

Flieger J, Swieboda R, Tatarczak M (2007) Chemometric analysis of retention data from salting-out thin-layer chromatography in relation to structural parameters and biological activity of chosen sulphonamides. J Chromatogr B 846(1–2):334–340

Frolova AM, Konovalova OY, Loginova LP et al (2011) Thin-layer chromatographic plates with monolithic layer of silica: production, physical–chemical characteristics, separation capabilities. J Sep Sci 34:2352–2361

Gandhi DB, Mehta PJ (2011) Simultaneous RP-HPTLC method for determination of Levodopa, Carbidopa, and Entacapone in combined tablet dosage form. J Planar Chromatogr -Mod TLC 24(3):236–241

Giaginis C, Tsantili-Kakoulidou A (2008) Alternative measures of lipophilicity: from octanol-water partitioning to IAM retention. J Pharm Sci 97(8):2984–3004

Greibrokk T, Andersen T (2003) High-temperature liquid chromatography. J Chromatogr A 1000:743–755

Grover PK, Ryall RL (2005) Critical appraisal of salting-out and its implications for chemical and biological sciences. Chem Rev 105(1):2–10

De la Guardia M, Khalaf KD, Carbonell V et al (1995) Clean analytical method for the determination of propoxur. Anal Chim Acta 308:462–468

De la Guardia M, Ruzicka J (1995) Towards environmentally conscientious analytical chemistry through miniaturization, containment and reagent replacement. Analyst 120:17N

Guillarme D, Heinisch S, Rocca JL (2004) Effect of temperature in reversed phase liquid chromatography. J Chromatogr A 1052:39–51

Gupta S, Shanker K, Srivastava SK (2012) HPTLC method for the simultaneous determination of four indole alkaloids in Rauwolfia tetraphylla: a study of organic/green solvent and continuous/pulse sonication. J Pharm Biomed Anal 66:33–39

Hartonen K, Riekkola M (2008) Liquid chromatography at elevated temperatures with pure water as the mobile phase. Trends Anal Chem 27:1–14

Jandera P (2008) Stationary phases for hydrophilic interaction chromatography, their characterization and implementation into multidimensional chromatography concepts. J Sep Sci 31:1421–1437

Jevric LR, Koprivtca GB, Misljenovic NM et al (2010) Chromatographic behavior and lipophilicity of s-triazine derivatives on silica gel impregnated with paraffin oil. Acta Period Technol 41:159–168

Jim SR, Oko AJ, Taschuk MT et al (2011) Morphological modification of nanostructured ultrathin-layer chromatography stationary phases. J Chromatogr A 1218:7203–7210

Kirchner JG (1967) Thin layer chromatography. In: Perry ES, Weissberger A (eds) Techniques of organic chemistry, 1st edn. Wiley, New York

Kochana J, Parczewski A, Wilamowski J (2006) SPE/TLC profiling of the impurities of MDMA: the influence of an agglutinant, diluents, and adulterants. J Liq Chromatogr Related Technol 29:1247–1256

Kowalska T, Kaczmarski K, Prus W (2003) Theory and mechanism of thin-layer chromatography. In: Sherma J, Fried B (eds) Handbook of thin-layer chromatography, 3rd edn. Marcel Dekker, New York-Basel

Krzek J, Starek M, Jelonkiewicz D (2005) RP-TLC determination of S(+) and R(–) ibuprofen in drugs with the application of chiral mobile phase and UV densitometric detection. Chromatographia 62:653–657

Kutsch H, Schoen U (2000) A simplified HPTLC screening method for the estimation of the PAH content in soil samples. Fresenius J Anal Chem 367:279–283

Lin M-C, Tsai M-J, Wen K-C (1999) Supercritical fluid extraction of flavonoids from Scutellariae Radix. J Chromatogr A 830:387–395

Malik I, Sedlarova E, Andriamainty F et al (2011) Synthesis and basic physicochemical properties of 1-[3-(Y-alkoxyphenylcarbamoyloxy)-2-hydroxypropyl]-4-(2-methylphenyl)piperazine chlorides. Ceska a Slovenska Farmacie 60(3):137–143

Mecozzi M, Amici M, Romanelli G et al (2002) Ultrasound extraction and thin layer chromatography–flame ionization detection analysis of the lipid fraction in marine mucilage samples. J Chromatogr A 963:363–373

Miller JM (2005) Chromatography: concepts and contrasts. Wiley, New York

Milojković-Opsenica DM, Malinar MJ, Tešić ŽLj (1999) Effect of the chelate ring size of diamine–Co(III) complexes on their hR_F values obtained under salting-out thin-layer chromatography conditions. J Chromatogr A 847:291–296

Miszczyk M, Pyka A (2006) Comparison of normal and reversed-phase TLC for separation of selected pesticides. J Planar Chromatogr 19:15–20

Mohammad A, Laeeq S (2011) Identification of coexisting pentose, hexose, and disaccharides with preliminary separation through hydrophilic interaction on silica HPTLC plate using aqueous sodium deoxycholate–acetonitrile mobile phase system. J Planar Chromatogr—Mod TLC 24 (6):491–496

Morlock G, Schwack W (2008) Planar chromatography—back to the future? LC-GC Eur 21:366–371

Nascu-Briciu RD, Sarbu C (2010) The lipophilicity determination of some pesticides by high performance thin-layer chromatography and various computing methods. Studia Universitatis Babes-Bolyai, Chemia 55(3):105–117

Odović JV, Stojimirović BB, Aleksić MB et al (2005) Examination of the Hydrophobicity of ACE inhibitors and their active metabolites by salting-out thin-layer chromatography. J Planar Chromatogr—Mod TLC 18(102):98–103

Oellig C, Schwack W (2011) Planar solid phase extraction—a new clean-up concept in multiresidue analysis of pesticides by liquid chromatography–mass spectrometry. J Chromatogr A 1218:6540–6547

Phattanawasin P, Sotanaphun U, Sukwattanasinit T et al (2012) Quantitative determination of sibutramine in adulterated herbal slimming formulations by TLC-image analysis method. Forensic Sci Int 219:96–100

Pobłocka-Olech L, Krauze-Baranowska M (2008) SPE-HPTLC of procyanidins from the barks of different species and clones of Salix. J Pharm Biomed Anal 48:965–968

Poole CF (2003) Thin-layer chromatography: challenges and opportunities. J Chromatogr A 1000:963–984

Poole SK, Poole CF (2011) High performance stationary phases for planar chromatography. J Chromatogr A 1218:2648–2660

Radoičić A, Majstorović H, Sabo T et al (2009) Hydrophilic-interaction planar chromatography of some water-soluble Co(III)-complexes on different adsorbents. J Planar Chromatogr—Mod TLC 22 (4):249–253

Rezić I, Horvat AJM, Babić S et al (2005) Determination of pesticides in honey by ultrasonic solvent extraction and thin-layer chromatography. Ultrason Sonochem 12:477–481

Ribeiro RLV, Bottoli CB, Collins KE et al (2004) Re-evaluation of ethanol as organic modifier for use in HPLC-RP mobile phases. J Brazil Chem Soc 15:300–306

Sarbu C, Djakovic-Sekulic T, Perisic-Janjic N (2002) Evaluation of lipophilicity of some benzimidazole and benzotriazole derivatives by RP HPTLC and PCA. J Pharm Biomed Anal 30:739–745

Šegan S, Andrić F, Radoičić A et al (2011) Correlation between structure, retention and activity of cholic acid derived cis-trans isomeric bis-steroidal tetraoxanes. J Sep Sci 34(19):2659–2667

Sharma UK, Sharma N, Gupta AP et al (2007) RP-HPTLC densitometric determination and validation of vanillin and related phenolic compounds in accelerated solvent extract of Vanilla planifolia. J Sep Sci 30:3174–3180

Sherma J (2003) Basic TLC techniques, materials, and apparatus. In: Sherma J, Fried B (eds) Handbook of thin-layer chromatography, 3rd edn. Marcel Dekker, New York-Basel

Shewiyo DH, Kaale E, Risha PG et al (2012) HPTLC methods to assay active ingredients in pharmaceutical formulations: a review of the method development and validation steps. J Pharm Biomed Anal 66:11–23
Shweshein AMSK, Radoičić A, Andrić F et al (2012) Hydrophilic interaction planar chromatography of geometrical isomers of some Co(III) complexes. J Liq Chromatogr Related Technol 35:1289–1297
Siouffi A-M (2005) From paper to planar: 60 years of thin layer chromatography. Sep Purif Rev 34:155–180
Skibinski R, Slawik T, Kaczkowska M (2011) Reversed-phase TLC study of the lipophilicity of fourteen 1,3-benzoxazol-2(3H)-one derivatives and comparison with isomeric 1,2-benzisoxazol-3(2H)-one analogs. J Planar Chromatogr—Mod TLC 24(4):348–351
Smith RM (2006) Superheated water: the ultimate green solvent for separation science. Anal Bioanal Chem 385:419–421
Stahl E (1956) Thin-Layer Chromatography. Pharmazie 11:633–637
Trifkovic J, Andric F, Ristivojevic P et al (2010) Structure-retention relationship study of arylpiperazines by linear multivariate modeling. J Sep Sci 33(17-18):2619–2628
Tsantili-Kakoulidou A (2005) Lipophilicity: assessment by reversed-phase TLC and HPLC. In: Encyclopedia of chromatography, 2nd edn, CRC Press. 993–998
Tuzimski T (2010) Application of HPLC and TLC with diode array detection after SPE to the determination of pesticides in water samples from the Zemborzycki reservoir (Lublin, Southeastern Poland). J AOAC Int 93(6):1748–1756
Vailaya A (2005) Fundamentals of reversed phase chromatography: thermodynamic and exothermodynamic treatment. J Liq Chromatogr Related Technol 28:965–1054
Vučković G, Živković-Radovanović V (2008) Use of different salt solutions in salting-out TLC of Co(III) complexes on silica gel. Chromatographia 67:259–267
Wang QS, Zhang L (1999) Review of research on quantitative structure-retention relationships in thin-layer chromatography. J Liq Chromatogr Related Technol 22(1):1–14
Welch CJ, Brkovic T, Schafer W et al (2009) Performance to burn? Re-evaluating the choice of acetonitrile as the platform solvent for analytical HPLC. Green Chem 11:1232–1238
Wu J, Lin L, Chao F (2001) Ultrasound assisted extraction of ginseng saponins from ginseng roots and ginseng cells. Ultrason Sonochem 8:347–352

Chapter 5
Green Techniques in Gas Chromatography

Dipa Biswas and Debarati Mitra

Abstract **Gas chromatography (GC)** is one of the most important analytical techniques among the various chromatographic processes currently in use. To widen it's applicability and acceptability analysts are now concentrating to develop green GC, replacing conventional GC. The appreciating feature of green GC is it's environment-friendliness by the way of reducing/eliminating the amount of solvents required for sample preparation and amount of waste generation or emission of volatile products. Reduction in chromatographic runtime and the possibilities of integrating GC with other efficient analytical tools are the other advantages of green GC which make it a highly efficient, sensitive and fast method of analysis in chemical science. This chapter highlights the different aspects of gas chromatography in the light of green techniques starting from sample preparation to the selection of mobile phase as well as chromatographic columns to be adopted. Coupling other analytical tools with GC to focus the versatility and high accuracy of analysis with dual system of separation and detection is also discussed.

5.1 Introduction

Gas chromatography (GC) also known as gas-liquid chromatography (GLC) is one of the most common and essential analytical tools for the separation and analysis of organic and inorganic mixtures. This separation technique is based on the partition equilibrium of an analyte between a liquid stationary phase and a gaseous mobile phase. Archer J. P. Martin and Richard L. M. Synge laid the foundation of GC in 1945 (Koel and Kaljurand 2010). The principal advantages associated with this technique are high process efficiency, strong separation power of even complex

D. Biswas (✉) · D. Mitra
Department of Chemical Technology,
92, A P C Road, 700009 Kolkata, India
e-mail: dipa_bis@yahoo.com

D. Mitra
e-mail: debarati_che@yahoo.com

Inamuddin, A. Mohammad (eds.), *Green Chromatographic Techniques*,
DOI 10.1007/978-94-007-7735-4_5, © Springer Science+Business Media Dordrecht 2014

mixtures, high sensitivity with small sample size, good precision and accuracy, relatively simple operation as well as low cost and long life of the instrument.

Over the past few years "going green" has become the mainstay of chemists, engineers and technologists. This is because apart from being beneficial, the activities of chemists and chemical engineers, both in laboratories and in industries, have adversely affected the global environment. Even, the activities of analytical chemists in the laboratories can exert a negative influence on the environment through uncontrolled disposal of chemical wastes. Adapting the principles of 'green chemistry' (Anastas 1998) shall definitely help in reducing the negative impacts of chemical products and processes on human health and environment. Besides the environmental benefits that come from inheriting greener technologies, the push for sustainability has also been triggered by economic conditions and material shortages. Smaller, faster, micro-scale instruments, along with solvent-saving techniques, can help to drive the eco-friendly movement in analytical chemistry. Methods that use less solvent generate less waste and reduce the expensive process of hazardous waste removal. Additionally, smaller, solvent-saving, sensitive instruments require less overall runtime, further adding to the energy efficiency of a laboratory.

The field of GC should also comprise the development of green analytical chemistry. The basic challenges to GC in the field of green analytical chemistry is the wastes produced during experimentation. The principle of three Rs (Reduce, Replace, Recycle) of green analytical chemistry are thus adopted in the framework of the new paradigm for gas chromatography (Welch et al. 2010). Various approaches for implementation of the principles of green analytical chemistry in gas chromatographic methods (Wardencki and Namieśik 2002) have been presented by Wardencki and Namiesik. This includes:

- Utilization of so-called direct methods of analysis, as much as possible, which permit determination of analytes in a sample without any pretreatment or sample preparation.
- Reducing energy consumption i.e. reducing sample preparation time when direct chromatographic analysis is not possible.
- Eliminating or reducing the amount of solvent during sample preparation steps applied before final chromatographic analysis.
- Conducting all operations with solvents in hermetic systems.
- Reducing emission of vapors and gases, as well as generation of liquid and solid wastes during analysis.
- Reducing matrix interferences.
- Reducing chromatographic run time.
- Reducing the necessity of reanalysis.
- Integrating different analytical procedures with GC.

In the light of the above approaches, the important aspects necessary to shift towards green GC are discussed below in details; among these the sample preparation part is the most important.

5.2 Sample Preparation

Typical methods of sample preparation of different analytes for GC analysis usually involve liquid-liquid or liquid-solid extraction with an organic solvent or preparation of suitable sample derivatives. These methods are time consuming and expensive. Moreover, the disposal of these usually toxic solvents is expected to offer adverse impacts on human health and environment. Using direct methods of gas chromatographic analysis i.e. eliminating the step of sample preparation prior to gas chromatographic analysis or adopting solventless sample preparation techniques (Wardencki and Namieśik 2000) are recommended for the implementation of the principles of green chemistry.

5.2.1 Direct Methods Without Sample Preparation

The sample preparation step is considered to be the most polluting step of the analytical process as it may involve the use of toxic chemicals and/or volatile solvents. Solvents used in analytical chemistry should be highly pure (chromatography grade); their manufacture requires greater number of steps for purification thus involving higher energy input and cost than technical-grade solvents. Introduction of direct methods of GC analysis, without pretreatment or sample preparation, is in coherence with the objectives of green analytical chemistry since it effectively reduces the impact of handling chemicals in the industrial laboratories. These techniques are simple, sensitive, reliable and reproducible. They also reduce labor, energy consumption, cost and time of analysis.

The two major fully automated techniques used nowadays for biological samples are headspace sampling and direct specimen injection. Compared to direct injection method, the headspace technique is less sensitive and requires larger volumes of the biological specimen. However, the headspace technique depending on the liquid matrix used is prone to serious analytical errors due to variations in partitioning of ethanol between the gas and liquid phases. Direct injection obviates all the sample type discrepancies observed with headspace analysis. The only drawback of direct injection method is its possible polluting effect on the injection port, or precolumn injection syringe. Deactivated precolumns can be installed in front of the analytical column to prevent inorganic salts and organic non-volatile compounds from entering into the column. The deactivated column also serves as a retention gap, which provides space for the solvent to evaporate. An alternative approach of injecting a liquid sample into the column is the Programmed Temperature Vaporization (PTV). This technique prevents the entry of solvent into the column as a result of evaporation of the solvent in the injector (Tobiszewski and Namieśik 2012). Another way to overcome this problem is to use a glass liner in the injector with proper sorbent to retain the solvent, as was developed by Tangerman (Tangerman 1997).

Willeboordse et al. (Willeboordse et al. 1968) developed a direct gas chromatographic method for the analysis of diaminotoluene isomers. Mixtures of 2-3, 3-4, 2-4, 2-5 and 2-6 diaminotoluenes were readily and accurately determined by utilizing a mixed partitioning agent of Carbowax 20M and Siponate DS-10 on the same solid support. The proposed method was obviously favorable in terms of green analytical chemistry compared to the conventional NMR and Derivatized GC methods.

Traditional methods of determining sterols in lipids of vegetables and animals require a saponification process for the conversion of acyglycerols into fatty acid soaps. The non-saponified fraction is separated from the soap by organic solvent extraction and then can be analyzed using GC. A rapid, precise and simple method without such derivatization process for determination of free sterols in vegetable and animal oils was developed by Choong et al. (Choong and Lin 1999) thereby eliminating the use of solvents and chemicals. A direct injection GC method with a non-polar megabore column was used.

Direct gas chromatographic analysis of volatile fatty acids in biological samples without prior derivatization, has been done by preparing aqueous extracts of volatile fatty acids (Pionetti 1979). The application of the direct chromatographic techniques is a step forward in reducing the environmental impact of chromatographic analysis.

5.2.2 Solventless Sample Preparation Techniques

Sample preparation prior to final determination is necessary in several biological and environmental samples which are not suitable for direct GC analysis. The sample preparation techniques most widely used in GC involve the following steps (Wardencki and Namieśik 2000):

1. Isolation of the analytes from the original matrix and moving them to the secondary matrix of simpler composition with an optional enrichment of the analytes,
2. Release of analytes kept in an appropriate trap either in isolation and/ enrichment stage,
3. Removal of excess solvent by drying, purification and fractionation of the extracts.

Organic analytes are conventionally separated from matrices by extraction methods such as liquid–liquid, solid–liquid which usually involve use of organic solvents followed by clean-up and preconcentration steps. The operations involved in sample preparation are usually time-consuming, tedious and difficult to automate. Moreover, the organic solvents used in most cases are toxic and their disposal after use, have the potential of inducing harmful effects on human health and the environment. Hence, the sample preparation step is considered to be potentially the most polluting step of analysis. Under such circumstances special attention is being paid to sample preparation methods which ensure reduction of the amount of liquid solvents or even their complete elimination during the course of sample preparation, i.e. solventless sample preparation techniques. Furthermore, the number of operations and processes involved in the sample preparation steps should be kept at a minimum. These aspects are strongly associated with both ecotoxicological and economic aspects. As a result of the application of solventless sample preparation techniques, the emission of toxic organic solvents into the environment is avoided. Additionally, the cost of using expensive high purity solvents and the cost of recycling them can be eliminated (Wardencki and Namieśik 2000).

The solventless sample preparation methods for GC analysis can be classified as follows:

1. Solid-Phase Extraction
2. Vapor-Phase Extraction
3. Thermal Desorption
4. Membrane Extraction

5.2.2.1 Solid Phase Extraction

Solid phase extraction is mainly used for trace analysis. A combination of solid phase extraction and thermal desorption of analytes from the sorbent bed leads to total elimination of solvents (Wardencki and Namieśik 2000). Solid phase extraction techniques are based on the distribution equilibria between the sample matrix and sorptive materials where analytes are extracted from the matrix into the non-miscible extracting phase. Sorptive extraction techniques have the following advantages:

1. Avoid the problem of emulsion formation that may occur in case of liquid-liquid extraction,
2. Automation of the technique.

Sorptive materials or sorbents are a group of polymeric materials with a glass transition temperature below the temperature at which the material is used during sampling, storage and desorption processes. Above their glass transition temperature, polymeric materials assume a liquid-like state, with properties similar to those of the organic solvents. Sorbents are non-porous homogeneous materials in which analytes can actually dissolve. Extraction of analytes depends on the partition coefficients of solutes between the phases. Selection of suitable adsorbents is very critical as it depends on the sample matrix and on the compound to be collected.

The arrangement of the sorbents is such that the least volatile compounds are trapped on the weakest sorbent at the front end of the tube, and successively more volatile compounds are trapped by increasingly strong sorbents further down the tube. Desorption then takes place in the reverse direction, as with single-bed tubes.

Six solid phase extraction techniques listed below, have been distinguished (Urbanowicz et al. 2011)

1. Open-tubular trapping (OTT)
2. Solid phase microextraction (SPME)

3. Stir bar sorptive extraction (SBSE)
4. Gum-phase extraction (GPE)
5. Closed loop stripping analysis (CLSA)
6. Sorption tubes

5.2.2.2 Vapor-Phase Extraction

The headspace (HS) analysis method is based on the partition law which states that, at given conditions of pressure and temperature, the ratio of the component concentrations in the liquid phase (L) and gas phase (G) at thermodynamic equilibrium is constant.

$$\frac{L}{G} = K\ (constant) \tag{5.1}$$

This ratio is called partition coefficient. Headspace analysis involves the transfer of the analytes from the original sample—condensed matter, mostly liquid—to the gas phase, which is to be analysed. Therefore, by combining headspace with an appropriate separation technique, it is possible to obtain information on the composition of the original sample (liquid or solid) based on analysis of the gas phase remaining in equilibrium with it.

Traditionally HS sampling operates either in static (S-HS) or dynamic mode (D-HS).

The different headspace techniques include (Urbanowicz et al. 2011)) (a) high concentration capacity headspace techniques (HCC-HS),(b) headspace solid-phase microextraction (HS-SPME),(c) headspace sorptive extraction (HSSE), (d) solid-phase aroma concentrate extraction (SPACE), (e) headspace solid phase dynamic extraction (HS-SPDE) and(f) multiple headspace extraction (MHSE).

5.2.2.3 Thermal Desorption (TD)/Thermal Extraction (TE)

Thermal desorption (extraction) being an attractive alternative to solvent extraction, is a widely used technique for extracting and isolating semivolatile compounds from various matrices. This method is well established in environmental analysis, food analysis and forensic science. In the process of TD, heat and inert gas (usually helium) are used to extract analytes retained in a sample matrix or on a sorbent bed. A temperature is needed that is high enough to allow desorption of the analytes from the matrix but also low enough to avoid degradation of the sample matrix itself. The analytes are desorbed into the gas stream and are ultimately transferred to the analyzer. TD in its most simple single-stage form is of limited application for packed column chromatography and cannot be used at all for capillary column GC. For this reason, most thermal desorbers are two-stage, i.e. they employ either a capillary cryo-focusing or cold adsorption trapping mechanism for concentrating analytes desorbed from the matrix before releasing them into the analytical system in as small a volume of vapor as possible. Both procedures do produce excellent, capillary-compatible chromatography, but capillary cryofocusing is quite costly and the volatility range of capillary cryofocusing devices is limited. Thermal desorption as a method of releasing organic compounds from the sorbent bed, offers the following advantages, in comparison with conventional solvent extraction (Czaplicka and Klejnowski 2002):

1. Typically 1,000 times more sensitive
2. Minimal sample preparation (eliminates the problem of contamination)
3. A smaller sample amount is required for the analysis
4. No analytical interference from solvent
5. Time efficiency
6. Greater than 90 % desorption efficiency
7. Selective focusing/extraction
8. Environmentally friendly (no solvent disposal)
9. Cost effectiveness
10. Eliminates problem associated with accurate dosing and repetition of the injection of liquid extracts
11. Eliminates the appearance of the solvent peak in the chromatogram, the components of which may mask the analytes
12. Eliminates difficulties associated with the choice of suitable solvents for the extraction of analytes especially when analytes differ significantly in polarity
13. The method is fully automated.

Thermal extraction is not without its limitations, because all types of substrates are not suitable for high temperature desorption. The use of TE is therefore complicated by the potential for carry-over, transfer loss, molecular rearrangement, fragmentation or breakdown of more thermally labile analytes at higher extraction temperatures and matrix effects, leading to quantification inaccuracies (Tsysik et al. 2008). Another drawback is sample consumption in a single analysis, although modern equipment incorporates design modifications to allow re-collection of split samples in a fresh tube.

The different thermal desorption methods are:

1. Direct thermal desorption (DTD)
2. Short-path thermal desorption (SPTD)
3. Temperature-programmed desorption (TPD)

5.2.2.4 Membrane Extraction

Several membrane materials in various configurations are presently being used for the extraction of a broad range of analytes from gaseous and aqueous matrices. Permeation through the membrane is dependent on the dissolution of the analyte in the membrane material i.e. the partition coefficient of a given analyte between the sample matrix and the membrane. The main advantage of the membrane technique is that the non-polar character of the membrane material allows minimal interference of water from aqueous samples. However, on the other hand, the low polarity of the membrane renders it unsuitable for the analysis of polar components (Wardencki and Namieśik 2000).

Suitable membrane devices such as flat film membranes or hollow fiber membranes are contacted on one side with the gaseous or liquid sample, while flushing the other side of the membrane with a stream of suitable gas. The resulting gas is directed straight to the measuring equipment.

The different modes of membrane extraction for isolating analytes from a stream of gas or liquid include:

1. Hollow fiber sampling analysis (HFSA) which is analogous to closed loop stripping analysis.
2. Membrane purge and trap (MPT)
3. Thermal membrane desorption (TMD)

Table 5.1 enumerates the main advantages and disadvantages of the different solventless sample preparation techniques mentioned above and the target compounds separated by the respective technique.

5.2.3 Sample Preparation Using Environmentally Friendly Solvents

Efforts have been made towards the development of procedures based on traditional, environmental friendly solvents. These techniques include:

1. Supercritical Fluid Extraction (SFE),
2. Subcritical Water Extraction (SWE),
3. Ionic Liquids (ILs).
4. Cloud-point Extraction.

5.2.3.1 Supercritical Fluid Extraction (SFE)

The superior solvation qualities of supercritical fluids over liquids enable SFE to extract organic pollutants from environmental samples in a short time. SFE uses non-toxic solvents, the most popular being CO_2, because (a) it has low critical parameters, (b) easily available at high purity, (c) capable to dissolve a wide spectrum of organics and (d) relatively cheap.

5.2.3.2 Subcritical Water Extraction (SWE)

In this technique pressurized hot water is used for the extraction because water is an eco-friendly, non-toxic, non-flammable, easy available and cheap solvent. The non-polar hydrocarbons are extracted by altering the dielectric constant of water.

Table 5.1 Advantages and disadvantages of the different solventless sample preparation techniques and the target compounds separated by the respective technique

Sample preparation technique	Advantages	Disadvantages	Target compounds	Ref.
Open-tubular trapping (OTT)	Sample cross-contamination and degradation are minimized Complete removal of water from the trap Very volatile compounds can be enriched at ambient temperatures Overcomes mechanical stability problems inherent to conventional SPME fibres	Low retention power for the trapping of very polar compounds from aqueous samples Complex instrumental setup and unfavorable sampling conditions Long sampling time for large sample volumes	t-butyl alcohol, butyl acetate, toluene, xylenes	(Dudek and Wolska 2002)
Solid-phase microextraction (SPME)	Rapidity, simplicity, sensitivity Compatibility with analyte separation and detection by different systems Provides linear results for a wide concentration of analytes Small in size (convenient for designing portable devices)	Limited range of stationary phases Relatively low operating temperature Breakage of the fibre Stripping of coatings Bending of the needle (expensive) Batch to batch variation of fibre coatings Robustness of fibre coatings Calibration must be made with the same solutions and/or samples.	H_2S, RSH, Me_2S	(Lestremau and Annika 2003)
In-needle solid phase microextraction	High sorption capacity No bleeding from thick-film coatings	Limited range of stationary phases	–	
Inside needle capillary adsorption trap (INCAT)	Simple methodology and easiness; rapidity of the analysis	Competitive effects and variation in sampling efficiencies for the analytes Low capacity of the sorbent Elution zones of analytes are slightly dispersed	Volatile organic compounds (BTEX)	(Kubinec and Berezkin 2004)

Table 5.1 (continued)

Sample preparation technique	Advantages	Disadvantages	Target compounds	Ref.
Needle trap devices (NTD)	Needle-like devices are particularly convenient for automation and development of on-line procedures In-expensive, robust and reusable	Limited range of stationary phases	Volatile organic compounds	(Wang et al. 2005)
Solid-phase dynamic extraction (SPDE)	Sensitivity is better and competition effects are largely eliminated compared to SPME High concentration factors Easy to apply and automate	Limited range of stationary phases Desorption temperature is limited by GC injection port	Pesticides	(Lipinski 2000)
Stir bar sorptive extraction (SBSE)	High enrichment factor compared to SPME	Limited range of stationary phases Specially designed thermal desorption units involve use of relatively sophisticated instrumentation Manual transfer of stir bar to the desorption unit may cause partial loss of the sensitivity gained	Benzophenone, 5-Chloro-2-(2,4-dicholorophenoxy) phenol, Organochlorine pesticides, carbamates, 4-Hydroxy buteric acid lactone	(Kawaguchi 2008; Tateo and Bononi 2003)
Gum-phase extraction (GPE)	Straightforward, simple and matrix-independent calibration Increase in sensitivity can occur, compared with OTT trap and an SPME fibre	Often fails for weakly retained solutes	Volatile organic compounds	(Baltussen and Davod 1998)
Equilibrium Gum-phase extraction (EGPE)	Achieved a higher sensitivity for all compounds than GPE	Expense of the more complicated calibration		
Closed-loop stripping analysis (CLSA)	Rapid and simple method for VOCs determination	Reproducibility problems CLSA is also not able to avoid foaming problems during analysis of waters with higher surfactant concentrations	Volatile organic compounds	(Martí and Lloret 2005)

Table 5.1 (continued)

Sample preparation technique	Advantages	Disadvantages	Target compounds	Ref.
Sorption tubes	High sampling versatility High VOC concentration power Easy portability Low cost and easy storage	Sorption and desorption efficiencies may not be 100 % Background impurities in sorbent tubes may interfere with analytes Limitations in sampling volume, rate, time Influence of sampling conditions on efficiency of sorption of analytes	Volatile organic compounds [16]	(Hollender and Sandner 2002)
Static headspace (S-HS)	Easy to use Volatile compounds in almost any sample matrix can be extracted Sample extraction, clean-up and preconcentration steps are not necessary	Lack of sensitivity Determination of trace compounds of relatively large molecular weight is very limited	Volatile compounds	(Lambropoulou et al. 2007)
Dynamic headspace (D-HS)	Highly reproducible and automated extraction procedure	Adsorption of water on trap	Chloroflurocarbons (CFCs)	(Massolo et al. 2009)
Headspace solid-phase microextraction (HS-SPME)	Simplicity, portability and low cost Shortens the time of extraction Facilitates analysis of solid samples	Limited range of stationary phases	Cannabinoids	(Gentili et al. 2004)
Headspace sorptive extraction (HSSE)	Amount of analyte presented in the headspace sampled by the twister can be easily determined Can be used to sample headspace with unfavorable β values and/or large headspace volumes High concentration capability	Need for dedicated and expensive instrumentations Limited range of stationary phases	Mercury and tin organometallics	(Prieto and Zuloaga 2008)
Solid-phase aroma concentrate extraction (SPACE)	Sensitive analysis of volatile analytes Good reproducibility	Not suitable for highly polar compounds		

Table 5.1 (continued)

Sample preparation technique	Advantages	Disadvantages	Target compounds	Ref.
Head-space solid-phase dynamic extraction (HS-SPDE)	Fully automated systems		Volatile organic compounds	(Jochmann et al. 2006)
Direct thermal desorption (DTD)	Small sample size Cost-effective sampling procedure Automated operation Short analysis time	Applicable only if the desired extraction takes place at a temperature below the decomposition point of materials in the sample matrix.	Volatile organic compounds	(Sanz et al. 2004)
Short path thermal desorption (STPD)	High sensitivity Minimum artifacts Losses and carryover effects		Volatile organic compounds	(Kuntasal and Karman 2005)
Temperature programmed desorption (TPD)	Complete transfer of high boiling analytes Avoids cross-contamination Fast analysis times		Plasticizers	(Wahl and Hoffmann 1999)
Thermal membrane desorption (TMD)	High sample throughputs High analysis cycle rate Good sensitivity	Complex set-up	Volatile and semi-volatile organic compounds	(Matz and Kibelka 1999)

5.2.3.3 Ionic Liquids (ILs)

Ionic liquids are non-aqueous, liquid or molten organic salts, being widely used to separate analytes like organic compounds and metallic ions. They are non-volatile, non-flammable and are considered as green solvents.

5.2.3.4 Cloud-Point Extraction

This technique is used to isolate organic and inorganic metallic species from aqueous solutions using a non-volatile, non-flammable and non-toxic surfactant.

5.2.4 Assisted Solvent Extraction

These methods of extraction are:

1. Microwave-assisted extraction (MAE), extraction based on absorption of microwave energy by molecules of chemical compounds.
2. Ultrasound-assisted extraction (UAE), in which extremely high local temperature and pressure gradients are employed for extracting analytes from solid matrices.
3. Pressurized liquid extraction (PLE), conventional liquid solvents are used at elevated pressures and temperatures for extraction.

Finally, it should be noted that there is no universal sample preparation technique, suitable for all types of samples. Sample preparation is dependent on the nature of the analytes present in the sample, the matrix, the final separation method and the type of information which is sought. However, it is most important that the sample preparation technique must be 'green'. Introducing such techniques in the laboratories in everyday practice will ensure to diminish the negative impacts of analytical chemistry on the environment.

5.3 Column Considerations for Green Gas Chromatography

Gas chromatographic columns are of two designs: packed and capillary. Packed columns are typically a glass or stainless steel coil (typically 1–5 m total length and 5 mm inner diameter) that is filled either with the stationary phase, or an inert packing coated with the stationary phase. Capillary columns, on the other hand are a thin fused-silica (purified silicate glass) capillary (typically 10–100 m in length and 250 μm inner diameter) that has the stationary phase coated on the inner surface.

Reduction of analysis time in GC and, consequently, increasing sample throughput contributes largely to green analytical chemistry. A fast separation can easily be achieved by using a short column packed with smaller particles, since column efficiency per unit length increases with decreasing particle size. The column length, L, required for a given separation is proportional to the particle size in case of packed columns (Giddings 1991):

$$L = 2N_e \times d_p \tag{5.2}$$

where N_e is the column efficiency for a given separation and d_p is the particle size of the packing material. Thus, solvent savings can be achieved by reducing the column length and particle size.

Again, analysis time in GC can also be reduced while maintaining separation and resolution by using columns with smaller i.d. (internal diameter, d_c). Plate number and column efficiency in GC depends on the capillary column i.d. ($N_e = L/d_c$). Reducing the analysis time by reduction of the characteristic diameter is accompanied by a requirement of higher inlet pressure. High resistance to flow of mobile phase in packed columns seriously limits the use of packed columns for fast GC. For fast GC, the columns should be operated near the conditions for minimum plate height. Narrow bore capillary columns with more plates per meter and having the same resolution can therefore be obtained using a shorter column length. Moreover, the optimum linear velocity of the carrier gas is increased, creating turbulent flow which also contributes to a further reduction of analysis time. Nevertheless, because of the high speed, good resolution and green nature, the capillary columns have been most popular in gas chromatography. Applying fast temperature programming by reducing the thermal mass of the column thermostat and using vacuum column outlet conditions (Cramers and Janssen 1999) are also some of the ways of increasing the speed of analysis leading to greening of the method.

Recently, rapid separations have been realized using multi-capillary columns (Pereiro et al. 1997; Mondello and Tranchida 2004) consisting of a parallel configuration of 900–2000 coated microcapillaries and i.d. of 20–40 μm. Such a bundle serves to eliminate the deficiencies associated with the capillary and packed columns while retaining the advantages of both. These columns enable fast separation of large volumes of samples injected quite efficiently.

5.4 Carrier Gas Considerations for Green Gas Chromatography

Carrier gas selection exerts a profound influence on both column performance and speed of analysis. Typical carrier gases (mobile phase) used in GC include helium, nitrogen and hydrogen.

As mentioned earlier, reduction in overall analysis time is one of the important principles of green analytical chemistry. The flow of the carrier gas through the GC column determines how fast an analysis can be accomplished. The mobile phase consumption (V) for a given separation can be expressed as (Giddings 1991):

$$V = F \times t \tag{5.3}$$

where F is the volumetric flow rate and t is the run time of separation. At a given temperature, the analysis time can be reduced by doubling the carrier gas velocity. Gas flows through the GC column when a pressure drop exists across the two ends—from the higher (inlet) to the lower (outlet) pressure region. The pressure-drop required to attain a specific carrier gas velocity is related to column length, column diameter and gas viscosity. Longer and narrower columns or higher gas viscosities will require a higher pressure drop to attain a given average gas velocity. Moreover, gas viscosities increase with temperature which has a marked influence on the performance of GC. Among the three gases, hydrogen has the lowest viscosity at any temperature, so it will produce higher velocities at a given pressure drop. On the other hand, helium and nitrogen have similar viscosities and they exhibit a steeper dependency upon temperature compared to hydrogen. Hence under identical conditions, hydrogen will allow the last peak to be eluted sooner compared to that of nitrogen and helium, thereby increasing the speed of analysis. Again, faster elution time allows lowering the maximum column temperature required for the analysis, which in turn reduces energy consumption. Lower temperatures also lead to less column bleeding and prolonged column life.

Narrow-bore capillary columns which are considered to be more eco-friendly compared to packed columns pose a greater pneumatic challenge. This is due to the fact that longer and narrower columns, with higher gas viscosities, require a higher pressure drop to yield a given average gas velocity. A 25 m × 100 μm column, for example needs more than 100 psig of helium to achieve a 35 cm/s average velocity at 50 °C. The dependence of pressure drop upon the square of the column diameter causes the gas velocity to increase dramatically. In this case, switching to hydrogen will reduce the pressure requirements to 43 psig at 50 °C and allow operations upto 350 °C at pressures below 100 psig. Nitrogen is the least preferred carrier gas for capillary columns because it has the most restricted operating range near the optimum linear velocity.

Helium is the most common carrier gas as it provides good separations, is inert and safe and can work with a greater number of detectors. However, in recent years the availability of high-purity helium has decreased and its cost has increased significantly. Moreover, helium, is non-renewable; it is a minor component of natural gas and is produced as a by-product of fractional distillation of natural gas and there are environmental concerns relating to the production and purification of the gas. In contrast to helium, hydrogen is considered to be a greener gas, since its production from water does not contribute to environmental pollution and is also readily available in high purity (Bartram and Froehlich 2010).

5.5 Coupling GC with Other Analytical Tools

For developing quick and organic solvent saving green procedures of analysis, GC is often coupled with other analytical tools like mass spectrometry (MS), fourier transform infra-red spectroscopy (FTIR), nuclear magnetic resonance (NMR), inductively-coupled plasma-MS (ICP-MS), atomic absorption spectroscopy (AAS) and atomic emission spectroscopy (AES). Recent developments in hyphenated techniques include GC-Olfactometry (GC-O) and multidimensional GC (MDGC or GC × GC). The aim of such coupling obviously helps to obtain an information-rich detection for both identification and quantification compared to that of a single detector.

GC-MS is a powerful separation and identification technique particularly in quality control applications. An unknown component can be identified quickly, without using a pure sample or standard, by applying GC-MS rather than using either of the single techniques. Zou et al. (Zou and Li 2006) developed a simple and environment friendly procedure for the determination of caffeine in beverages by using GC-MS in which the total amount of organic solvent consumed is only 1 ml. The advantage of comprehensive two dimensional GC coupled with MS (GC × GC-MS) is the increased speed of analysis particularly in case of complex mixtures.

Normally, the NMR analysis of organic compounds of synthetic or natural products, which are often not available in the pure form, is performed in two independent steps: separation of the mixture into individual components followed by identification. Isolation of individual components involves the use of harmful organic solvents, energy and time consumption. An online combination with GC can eliminate the separation step and accelerate the analysis.

The widespread use of organometallic compounds in agriculture, plastics industry and subsequently their release into the environment has led to an increasing concern about their persistence and toxicity. For instance, organotins unlike inorganic tin, may occur at toxic levels in aquatic and sedimentary environments. The distribution of organotins as well as its bioavailability and toxicity depend critically on the chemical form in which the species are actually present. This is why the determination of total tin by AAS alone is not able to provide reliable information on the hazards of this element to human health and environment. Therefore, speciation analyses of these organometallic compounds are of great importance (Guo and Lankmayr 2012). Since its introduction, GC-AAS has been extensively applied to speciation of organometallic compounds such as Sn, Pb and As. Just because of this, several derivatization methods have been reported in order to convert them to volatile species for GC analysis and a sensitivity of picogram level can be achieved together with a highly selective detection as the result of the coupling. Furthermore, GC-AES provides a rapid method for sensitive screening of human urine samples without dilution (Zachariadis and Rosenberg 2009).

5.6 On-Site Analysis

The analytical procedure usually comprises of the following steps:

1. Sample collection,
2. Transportation and storage,

3. Sample preparation,
4. Chromatographic separation and detection.

There is always a chance of producing analytical wastes in each of these steps. However, by placing the GC in the form of in-line, on-line or at-line modes, the first two steps can be omitted. Hence, over the years, a trend towards on-site measurement has been observed. Several portable or transportable GC and GC–MS systems are now commercially available that allow on-site analysis. All these systems can be considered contributory to green analytical methods, because the expensive, energy-consuming transport of samples to laboratories is eliminated. On-site sampling also has the advantage that remediation or re-sampling can take place immediately after the first analytical results are obtained.

Recently, a transportable system combining a low thermal mass (LTM)–GC with a standard mass spectrometry detector (MSD) system has been introduced by Agilent Technologies (Agilent Technologies, 2010). Using this system, on-site analysis is possible following the same technology and achieving the same performance as available within a laboratory and sample injection can be performed by liquid split/splitless injection, gas sampling valve, static headspace or solid-phase microextraction (SPME).

5.7 Conclusion

Environmental-friendly practices in different areas of research lead to the development of green methodologies of analysis. They have a dramatic impact on the society and have proved to be beneficial from various aspects. Green analytical chemistry (GAC) was thus developed of which GREEN GC is one of the option. Progress in various analytical methodologies has contributed to the development of this new, greener option in the field of gas chromatography. The major development in gas chromatography as a green analytical tool has two main features comprising of sample preparation technique and fast analysis time. These techniques result in the use of smaller amounts of harmful solvents and reagents, reduce waste generation and require less energy. But additional costs are involved in changing over to a greener alternative technology; hence in switching over to green GC, it should be relevant to think of the techniques/ instruments/accessories which are both cheaper and greener. Though the implementation of this green technology in the field of gas chromatography lag behind but the future of this technology is clear and we can look forward for the day not too distant for this high-valued technique to be adopted in all the laboratories for sustainable development of both the environment & the society.

References

Agilent technologies introduces transportable GC/MS for High-Performance, Out-of-laboratory analysis.http://www.agilent.com/about/newsroom/presrel/2010/01mar-ca10015.htmlAccessed 1 June 2012

Anastas PT (1998) Green chemistry. Theory and practice. Oxford University Press, New York

Baltussen E, Davod F et al (1998) Soprtion tubes packed with polydimethylsiloxane: A new and promising technique for the preconcentration of the volatiles and semi-volatiles from air and gaseous samples. J High Resol Chromatogr.21:332–340

Bartram RJ, Froehlich P (2010) Considerations on switching from Helium to Hydrogen http://www.labgasgenerators.com/Docs/Articles/LCGC1010_He_to_Hydrogen_Article.pdf. Accessed 1 June 2012

Choong Y-M, Lin H-J et al (1999) A rapid gas chromatographic method for direct determination of free sterols in animal and vegetable fats and oils. J Food Drug Anal 7:279–290

Cramers CA, Janssen H-G et al (1999) High-speed gas chromatography: an overview of various concepts. J Chromatogr A 856:315–329

Czaplicka M, Klejnowski K (2002) Determination of volatile organic compounds in ambient air. J Chromatogr A 976:369–376

Dudek M, Wolska L et al (2002) The application of an open tubular trap in analysis of organic air pollutants. Chemosphere 48:913–918

Gentili S, Cornetta M, Macchia T (2004) Rapid screening procedure based on head space solid-phase microextraction and gas chromatography-mass spectrometry for the detection of many recreational drugs in hair. J Chromatogr B 801:289–296

Giddings JC (1991) Unified Separation Science. Wiley, New York

Guo X, Lankmayr E (2012) Hyphenated techniques in gas chromatography. Advanced gas chromatography—progress in agricultural, biomedical and industrial applications. doi:10.5772/31802

Hollender J, Sandner F et al (2002) Sensitive indoor air monitoring of monoterpenes using different adsorbents and thermal desorption gas chromatography with mass- selective detection. J Chromatogr A 962:175–181

Jochmann MA, Kmiecik MP, Schmidt TC (2006) Solid phase dynamic extraction for the enrichment of polar volatile organic compounds from water. J Chromatogr A 1115:208–216

Kawaguchi M, Ito R et al (2008a) Determination of urinary triclosan by stir bar sorptive extraction with in-situ derivatization and thermal desorption-gas chromatography-mass spectrometry. J Chromatogr B 875:577–580

Kawaguchi M, Ito R et al (2008b) Simultaneous analysis of benzophenone sunscreen compounds in water sample by stir bar sorptive extraction with in-situ derivatization and thermal desorption- gas chromatography- mass spectrometry. J Chromatogr A 1200:260–263

Koel M, Kaljurand M (2010) Green analytical chemistry. RSC Publishing, Cambridge

Kubinec R, Berezkin VG et al (2004) Needle concentrator for gas chromatoghraphic determination of BTEX in aqueous samples. J Chromatogr B 800:295–301

Kuntasal ÖO, Karman D et al (2005) Determination of volatile organic compounds in different micro environments by multi-bed adsorption and short-path thermal desorption followed by gas chromatographic mass spectrometric analysis. J Chromatogr A 1099:43–54

Lambropoulou DA, Konstantinoou IK, Albanis TA (2007) Recent developments in head-space microextraction techniques for the analysis of environmental contaminants in different matrices. J Chromatogr A 1152:70–96

Lestremau F, Annika F et al (2003) Evaluation of solid-phase microextraction for time-weighted average sampling of volatile sulfur compounds at ppb concentrations. Anal Chem 75: 2626–2632

Lipinski J (2000) Automated multiple solid phase micro extraction. An approach to enhance the limit of detection for the determination of pesticides in water. Fresenius J of. Anal Chem 367:445–449

Martí I, Lloret R et al (2005) Determination of chlorinated toluenes in raw and treated water samples from the Llobregat by closed loop stripping analysis and gas chromatography- mass spectrometry detection. J Chromatogr A 1077:68–73

Massolo S, Rivaro P, Frache R (2009) Simultaneous determination of CFC-11, CFC-12 and CFC-113 in seawater samples using a purge and trap gas chromatographic system. Talanta 80:959–966

Matz G, Kibelka G (1999) Experimental study on solventless sample preparation methods: Membrane extraction with a sorbent interphase, thermal membrane desorption application and purge-and-trap. J Chromatogr A 830:365–376

Mondello L, Tranchida PQ et al (2004) Fast GC analysis with a 50 μm ID column: Theory, practical aspects and application to a highly complex sample. J Sep Sci 27:1149–1156

Pereiro IR, Schmitt VO, Lobinski R (1997) Elemental speciation analysis by multicapillary gas chromatography with microwave-induced plasma atomic spectroscopic detection. Anal Chem 69:4799–4807

Pionetti JM (1979) Recovery of volatile fatty acids from biological materials for direct analysis by gas chromatography. J of Biochem Biophys Methods 1:71–76

Prieto A, Zuloaga O et al (2008) Simultaneous speciation of methyl mercury and butyl tin species in environmental samples by head space-stir bar sorptive extraction-thermal desorption-gas chromatography-mass spectrometry. J Chromatogr A 1185:130–138

Sanz J, Soria AC, Garcia-Vallejo MC (2004) Analysis of volatile components of *Lavandula luisieri L.* by direct thermal desorption-chromatography-mass spectrometry. J Chromatogr A 1024:139–146

Tangerman A (1997) Highly sensitive gas chromatographic analysis of ethanol in whole blood, serum, urine and fecal supernatants by the direct injection method. Clin Chem 43:1003–1009

Tateo F, Bononi M (2003) Determination of gamma-butyrolactone (GBL) in foods by SBSE-TD/GC/MS. J Food Compos Anal 16:721–727

Tobiszewski M, Namieśik J (2012) Direct chromatographic methods in the context of green analytical chemistry. Trends Anal Chem 35:67–73

Tsysik P, Czech J, Carleer R (2008) Thermal extraction coupled with gas chromatography-mass spectrometry as a tool for analyzing dioxin surrogates and precursors in fly ash. J Chromatogr A 1210:212–221

Urbanowicz M, Zabiegala B, Namieśik J (2011) Solventless sample preparation techniques based on solid- and vapor-phase extraction. Anal Bioanal Chem 399:277–300

Wahl HG, Hoffmann A et al (1999) Identification of plasticizers in medical products by a combined direct thermodesorption-cooled injection system and gas chromatography-mass spectrometry. J Chromatogr A 847:1–7

Wang A, Fang F, Pawliszyn J (2005) Sampling and determination of volatile organic compounds with needle trap devices. J Chromatogr A 1072:127–135

Wardencki W, Namieśik J (2000) Solventless sample preparation techniques in environmental analysis. J High Resol Chromatogr 23:297–303

Wardencki W, Namieśik J (2002) Some remarks on gas chromatographic challenges in the context of green analytical chemistry. Pol J Environ Stud 11:185–187

Welch CJ, Wu N, Biba M (2010) Greening analytical chromatography. Trends Anal Chem 29: 667–680

Willeboordse F, Quick Q, Bishop ET (1968) Direct gas chromatographic analysis of isomeric diaminotoluenes. Anal Chem 40:1455–1458

Zachariadis G, Rosenberg E (2009) Speciation of organotin compounds in urine by GC-MIPAED and GC-MS after ethylation and liquid-liquid extraction. J Chromatogr B 877:1140–1144

Zou J, Li N (2006) Simple and environmental friendly procedure for the gas chromatographic-mass spectrometric determination of caffeine in beverages. J Chromatogr A 1136:106–110

Chapter 6
Preparation and Purification of Garlic-Derived Organosulfur Compound Allicin by Green Methodologies

Jenny Lee, Shalini Gupta, Jin-Sheng Huang, Lasanthi P. Jayathilaka and Bao-Shiang Lee

Abstract Effective green chemistry, green reversed-phase high-performance liquid chromatography (RP-HPLC), and green thin layer chromatography (TLC) methods used for preparing and purifying allicin, a garlic-derived organosulfur compound, are described here. A greener version of the acidic oxidation reaction of diallyl disulfide (DADS) is used to produce allicin with high yield. Green RP-HPLC eliminates the liquid/liquid extraction step from either the DADS acidic oxidation reaction mixture or from garlic extract, allowing the single-step purification of allicin. The proposed method involved the use of eco-friendly ethanol as the alternative eluent for acetonitrile. The pure allicin HPLC fraction prepared this way is quite stable and can be used directly for chemical and biological applications. In addition, by changing silica gel TLC plate to RP-C_{18} TLC plate, 50 % aqueous ethanol, instead of a solvent blend of hexane : isopropanol (92:8) can be used to identify the allicin on TLC plate. Here, the traditional usage of toxic organic solvents has been avoided and a more efficient chemical reaction scheme is employed, which permits the classification of the present method as green. These green methodologies are used successfully to prepare pure allicin, investigate the thermal, pH, and vacuum drying decomposition of the allicin, and analyze various preparations of garlic extract.

6.1 Introduction

Allicin (3-prop-2-enylsulfinylsulfanylprop-1-ene), chemical structure is shown in Fig. 6.1, is one of the most biologically active compounds found in nature (Block 2010; Ali et al. 2000). It is a thioester of sulfenic acid or allylthiosulfinate derived

B.-S. Lee (✉) · S. Gupta · J.-S. Huang · L. P. Jayathilaka
Protein Research Laboratory, Research Resources Center, University of Illinois at Chicago, 835 S. Wolcott Avenue,
Chicago, IL 60612, USA
e-mail: boblee@uic.edu

J. Lee
Illinois Mathematics and Science Academy, 1500 Sullivan Rd,
Aurora, IL 60506-1000, USA

Inamuddin, A. Mohammad (eds.), *Green Chromatographic Techniques*,
DOI 10.1007/978-94-007-7735-4_6, © Springer Science+Business Media Dordrecht 2014

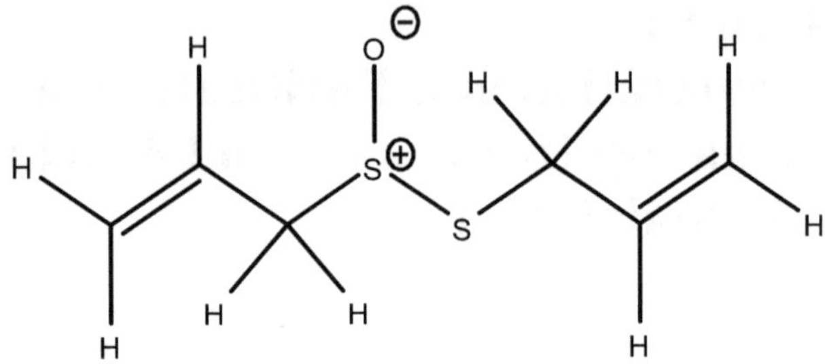

Fig. 6.1 Chemical structure of allicin

from garlic. Animal studies have reported that allicin reduces atherosclerosis (Block 2010; Gonen et al. 2005), decreases blood pressure (Block 2010; Ali et al. 2000), and acts as an antioxidant (Block 2010; Ilic et al. 2010). In addition, it exhibits anti-bacterial, anti-fungal, anti-viral, anti-cancer, anti-thrombotic and anti-inflammatory properties (Block 2010; Ali et al. 2000; Ankri and Mirelman 1999; Cutler and Wilson 2004; Oommen et al. 2004; Ariga and Seki 2006; Feng et al. 2012). Purified allicin is scarcely available and commercially expensive due to its instability. Allicin is readily decomposed into oil-soluble organosulfur compounds (OSCs) such as diallyl sulfide, diallyl disulfide (DADS), diallyl trisulfide, dithiins, and ajoene as well as to some water-soluble OSCs such as S-allyl cysteine and S-allyl mercaptocysteine (Lanzotti 2006; Block 1985; Ichikawa et al. 2006). Allicin can be prepared chemically (Ilic et al. 2010; Lee et al. 2013; Cremlyn 1996; Cruz-Villalon 2001), enzymatically (Miron et al. 2004; Shadkchan et al. 2004), and biologically (Lee et al. 2013; Yu and Wu 1989). The chemical synthesis of allicin is based on the oxidation of DADS. Enzymatic synthesis involves the chemical synthesis of alliin, and then its subsequent conversion to allicin using alliinase. Alliinase is produced by crushing garlic, which releases alliinase (which makes up 10 % w/w of total protein) from the bundle sheath cells. Pure biological synthesis is achieved from crushing the garlic cloves in several solvents and then extracting the allicin from garlic extract.

Previously, we developed efficient strategies to identify and purify allicin in aqueous garlic extract (AGE) or DADS acidic oxidation reaction by using RP-HPLC, TLC, MS, NMR, and MTT [3-(4,5-Dimethylthiazol-2-yl)-2,5-diphenyltetrazolium bromide] cell viability assay (Lee et al. 2013). In this method, pH of the AGE is changed from 7 to 5 to eliminate interfering molecules and to enable a clean RP-HPLC separation of the constituents in AGE. MTT assay of the HPLC fractions is used to identify the active fractions. Pure allicin fraction in HPLC solvent turns out to be quite stable and is ready to use. Further identification of allicin in the HPLC fraction is done by TLC, MS, and NMR. In response to the public outcry for an environmentally responsible practice of basic chemical activities (Guardia and Garrigues 2012), the major goal of the present work is to develop a greener version of the method to prepare and purify allicin from either AGE or DADS acidic oxidation reaction mixture. The strategy presented here reduces the use of organic solvents and maximizes the efficiency of chemical processes. The thermal, pH, and concentration stability data of the allicin are presented in this chapter. Allicin content in garlic extract prepared by different solvents or by stewing or stir frying the finely chopped garlic have been evaluated by green RP-HPLC. A greener allicin chemical synthesis is also described.

6.2 Green RP-HPLC Purification of the Allicin

Since it is necessary to separate and purify allicin from multi-component samples, the reversed-phase high-performance liquid chromatography (RP-HPLC) method is best suited for the job (Brown and Grushka 2001). However, RP-HPLC requires a large amount of hazardous organic solvents such as acetonitrile, methanol, tetrahydrofuran, or isopropanol as a mobile phase. The polar and water-miscible solvent acetonitrile is the most common eluent in RP-HPLC. In 2008, decreasing demand for acrylonitrile due to the financial crisis resulted in acetonitrile, a by-product of acrylonitrile production, worldwide shortage (Brettschneider et al. 2010). It was this shortage and the desire to reduce the expensive process of hazardous waste removal that pushed scientists toward using alternative, less toxic solvents such as ethanol. We have successfully developed such eco-friendly methods that avoid the use of hazardous reagents and keep the chemical waste to a minimal level. The method is based on the modification of a RP-C_{18} column, with ethanol instead of acetonitrile as HPLC solvent. The traditional organic solvent is avoided, which permits the classification of the present method as green. To simplify the detection of the active components in the AGE, majority of the proteins were eliminated by changing the pH of the AGE from 7 to 5. This is accomplished by adding 50 μL 1M HCl to 950 μL AGE and then centrifuging for 4 min at 10,000x g. The samples were separated by a gradient RP-HPLC with a Kinetex column (Phenomenex, C_{18}, 4.6 mm ID×250 mm L, 5 μm particle size,100 Å pore size) at 1 mL/min using a binary solvent system consisting of A=0.1% TFA (v/v) in water and B: 0.1% TFA (v/v) in ethanol at flow rate of 1 mL/min. A gradient was run by holding at 5% B for 5 min followed by 5–10% B in 2 min, 10–35% B in 8 min, and 35–100% B in 15 min. The retention time for allicin was 18.3 min at ambient temperature (allicin is detected at 250 nm). Pharmaceutical grade 200 proof anhydrous ethanol was used. In addition to being less expensive than either HPLC grade acetonitrile or methanol, it also has a higher flash point (14 °C), low volatility, and cheaper disposal costs (biodegradable). The UV cutoff and the residue absorbance of ethanol do not interfere with the detection of the analytes. Although the back pressure is higher than traditional method due to the higher viscosity of ethanol, it does not create problems. Figure 6.2 shows the traditional and green RP-HPLC of allicin from AGE and chemical synthesis. The chromatographic procedure yields precise results within 30 min with the use of 15 mL of ethanol. These results clearly show that the green RP-HPLC is comparable to the conventional method, which uses acetonitrile as mobile phase. In addition, ethanol can elute the highly retained analytes at the end of the gradient more effectively than acetonitrile or methanol. The

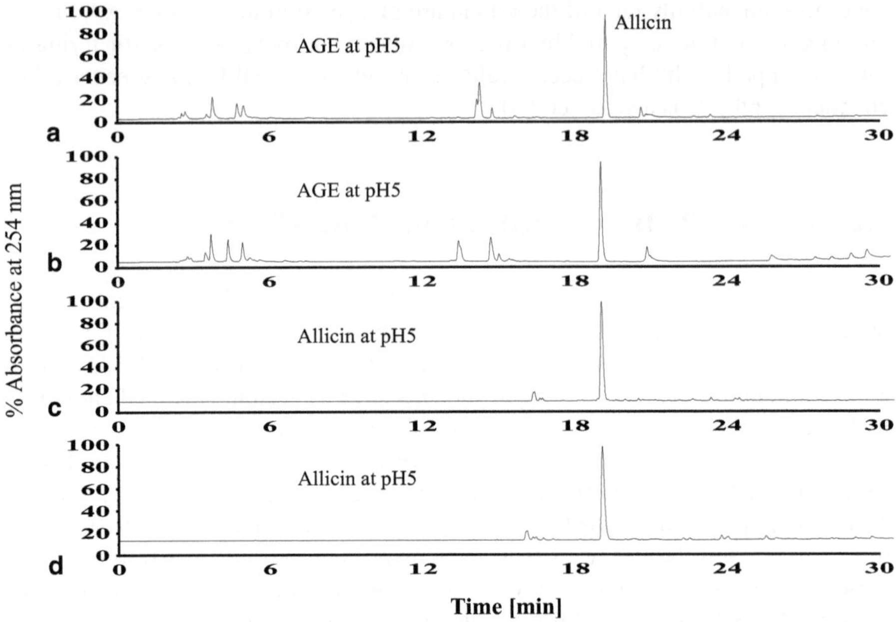

Fig. 6.2 Green and traditional RP-HPLC chromatograms of aqueous garlic extract (AGE) and pure allicin: (**a**) traditional RP-HPLC of AGE at pH 5. (**b**) Green RP-HPLC of AGE at pH 5. (**c**) Traditional RP-HPLC of allicin at pH 5. (**d**) Green RP-HPLC of allicin at pH 5. RP-HPLC was done under the following conditions: Instrument; Hewlett Packard 1100 (Agilent Technologies, Santa Clara, CA). Column; Kinetex RP-C_{18} (Phenomenex, Torrance, CA), 5 μm, 250 × 4.6 mm. Solvent A: 0.1 % TFA in H_2O. Solvent B: 0.1 % TFA in acetonitrile and 0.1 % TFA in ethanol for traditional and green RP-HPLC, respectively. Temperature: 30 °C. Flow rate: 1 mL/min. Elution: A gradient was run by holding at 5 % B for 5 min followed by 5–10 % B in 2 min, 10–35 % B in 8 min, and 35–100 % B in 15 min. Detection wavelength; 254 nm. Allicin is eluted at 18.3 min

active fraction at 18.3 min is identified as allicin by TLC, MS, and NMR and further confirmed using a pure synthetic allicin. It is interesting to know that after sample analysis, ethanol could be recycled and reused as an eco-friendly alternative fuel.

6.3 Characterization of the Allicin by Green Methodologies

For the identification of active fraction allicin, various techniques, such as TLC, MS, and ^{1}H-NMR were used (Block 2010; Ilic et al. 2010; Lee et al. 2013; Cruz-Villalon 2001; Macpherson et al. 2005; Fujisawa et al. 2008). The presence of allicin could be identified on aluminum backed silica gel F_{254} TLC plate (Dynamic Adsorbents, Norcross, GA). Pure allicin was used as a control for comparison. Hexane : isopropanol (92:8) was used as the mobile phase, compounds were detected at 254 nm, and the R_f value was calculated to be 0.44 (Figs. 6.3a, b showing the traditional

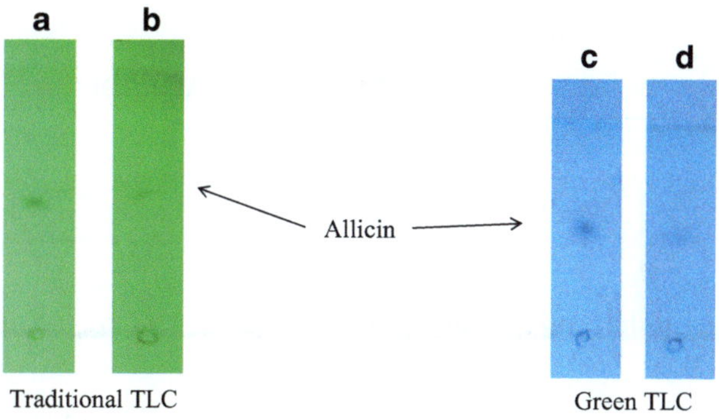

Fig. 6.3 Green and traditional TLC chromatograms of allicin purified from aqueous garlic extract (AGE) and DADS oxidation reaction: (**a**) traditional TLC of allicin from AGE. (**b**) Traditional TLC of allicin from DADS oxidation reaction. (**c**) Green TLC of allicin from AGE. (**d**) Green TLC of allicin from DADS oxidation reaction. Traditional TLC was done on 200 μm aluminum backed silica gel with F_{254} indicator TLC plate (Dynamic Adsorbents, Norcross, GA). Hexane : isopropanol (92:8) was used as the mobile phase and compounds were detected at 254 nm. R_f value of allicin was calculated to be 0.44. Green TLC was done on 200 μm aluminum backed reversed-phase C-18 stationary phase with F_{254} indicator TLC plate (Dynamic Adsorbents). The eco-friendly solvent system of ethanol : water (1:1) was used as the mobile phase and compounds were detected at 254 nm. R_f value of allicin was calculated to be 0.51

TLC of pure allicin from either AGE or chemical synthesis). To decrease organic waste, we used green TLC to identify the allicin on aluminum backed reversed-phase stationary phase (200 μm) with F_{254} indicator C_{18} TLC plate. Ethanol : water (1:1), an environmentally friendly solvent system, was used as the mobile phase. Compounds were detected under UV radiation (254 nm) and the R_f value was calculated to be 0.51 (Figs. 6.3c, d showing the green TLC spot of the pure allicin from either AGE or chemical synthesis). With plain silica TLC plates, a non-polar solvent was typically used as the mobile phase, thus pulling polar or ionic analytes from the non-polar phase as the solvent travels through the plate. Silica RP-C_{18} plates were composed of silica particles which have carbon chains (18 carbons long) attached to the silica matrix. When the non-polar C_{18} chains are attached to the normally polar/hydrophilic stationary phase, the silica matrix acts in the opposite manner and attracts non-polar analytes. Typically, a completely different mobile phase is used, usually one that is slightly polar and allows the mobile phase to travel without interacting with the plate. Solvents such as water, methanol, ethanol, acetonitrile, etc. can be used in developing the RP plate. Here, the water and ethanol are used with good results; however, green reversed-phase C_{18} TLC is more expensive and requires three times longer duration for the development of TLC plates.

Figures 6.4 and 6.5 show MS and NMR spectra of the pure allicin, respectively. MS (MALDI-TOF/TOF, CHCA as matrix, MH^{+1}, Fig. 6.4a; CID spectrum of 163, Fig. 6.4b; m/z 163). MS (ESI, MH^{+1}, Fig. 6.4c; CID spectrum of 163, Fig. 6.4d; m/z 163). ^1H NMR (Bruker 900 MHz, D_2O, δ /ppm, DSS as an internal standard, Fig. 6.5): δ 3.83 (1H, S-CH_2-C, dd, J=14, J=7), δ 3.86 (1H, S-CH_2-C, dd, J=14,

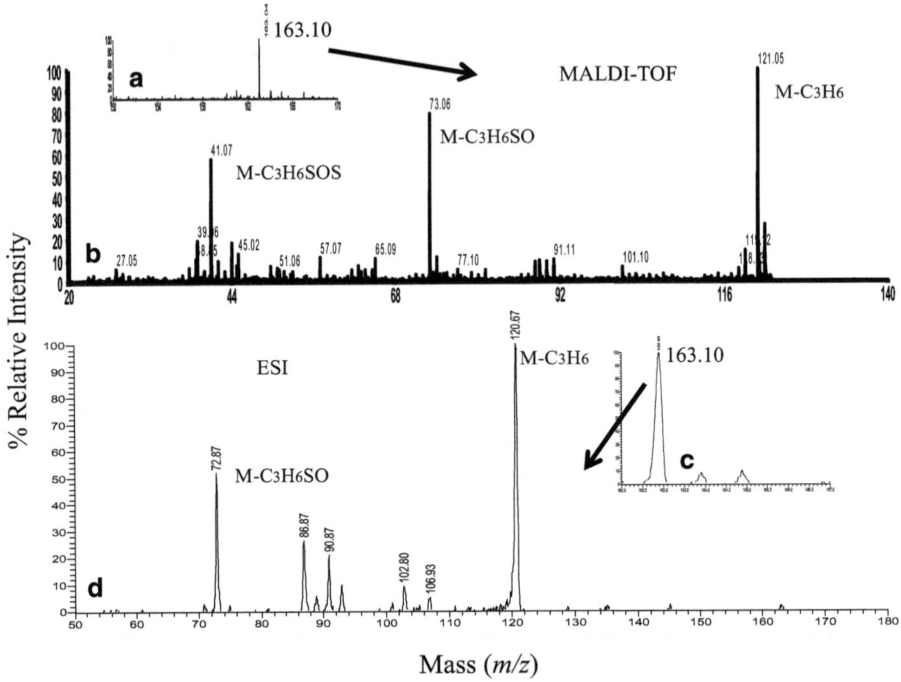

Fig. 6.4 Mass and collision induced dissociation (CID) spectra of allicin: (**a**) MALDI-TOF MS. (**b**) CID masss pectrum(MALDI-TOF/TOF) of *m/z* 163. (**c**) ESI-MS. (**d**) CID masss pectrum (ESI-MS) of *m/z* 163. The matrix used for MALDI is cyano-4-hydroxycinnamic acid. Methanol was used as the solvent for ESI. Both MALDI-TOF MS and ESI-MS are effective techniques to characterize allicin

J=7), δ 3.93 (1H, C-C**H**$_2$-SO, dd, J=13, J=7), δ 4.01 (1H, C-C**H**$_2$-SO, dd, J=13, J=7), δ 5.25 (1H, S-CH$_2$-CH=C**H**$_2$, d, J=10), δ 5.36 (1H, S-CH$_2$-CH=C**H**$_2$, d, J=17), δ 5.49 (1H, C**H**$_2$=CH-CH$_2$-SO, d, J=17), δ 5.52 (1H, C**H**$_2$=CH-CH$_2$-SO, d, J=10), δ 5.94-6.01 (2H, CH$_2$-C**H**-CH$_2$, ddt, J=17, J=10, J=7). The spin coupling pattern between δ 3.83 and δ 3.86 can be understood by non-first-order pattern resulting from the ratio between the chemical shift difference in Hz and the spin coupling of 2.2 (expressed as $\Delta v/J=31.3/14=2.2$). In addition, the spin coupling pattern between δ 3.93 and δ 4.01 can be understood by non-first-order pattern resulting from $\Delta v/J=74.6/13=5.7$ (Crews et al. 2009).

6.4 Allicin in Different Garlic Extract by Green RP-HPLC

Allicin can be obtained from garlic extract (GE) directly (Lee et al. 2013; Yu and Wu 1989). However, the content of allicin in GE can vary depending on the method of preparation and storage conditions (Kim et al. 2012; Islam et al. 2011; Song and

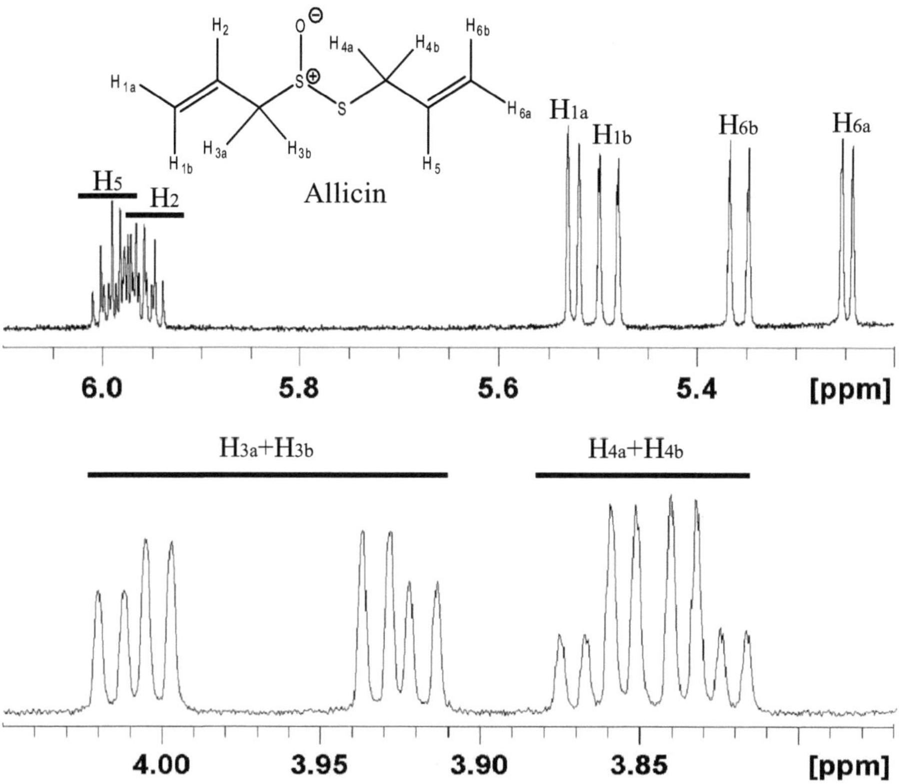

Fig. 6.5 One-dimensional 1HNMR spectrum (Bruker 900 MHz spectrometer) and peak assignments of allicin. Allicin was dissolved in D_2O and DSS (4,4-dimethyl-4-silapentane-1-sulfonic acid) was used as an internal calibration standard

Milner 2001). It is reported (Fujisawa et al. 2008) that ethanolic aqueous solvents with >20 % water were superior then water as extractants for allicin. On the other hand, vegetable oil and n-hexane were unable to extract allicin. We used the following procedure to prepare garlic extract to test the effectiveness of various solvents. Garlic (Spice World, Orlando, FL) was purchased from the local market, and the aqueous garlic extract was prepared from 23 g of skinned garlic cloves in 40 mL solvent. The mixture was crushed in a blender (NutriBullet, Los Angeles, CA) for 10 seconds at one pulse per second, left to rest for 10 min, and then crushed at one pulse per second for 10 additional seconds. The extract solution of the crushed garlic was filtered through a cheese cloth. The filtrate was aged overnight allowing any particulate to settle. On day two, the GE was centrifuged at 900 × g for 4 min, and then again at 10,000 × g for another 4 min. Finally, the supernatant was filtered through a sterilized 0.22 μm syringe-driven filter (Becton Drive, Franklin Lakes, NJ) to clean out any residual precipitate and bacteria. The clear supernatant was aliquoted and stored at −80 °C until used. To simplify the purification of allicin, majority of the proteins were eliminated by lowering the pH of the GE from 7 to 5. Solvents such as water, ethanol, DMSO, and hexane were used to make the garlic

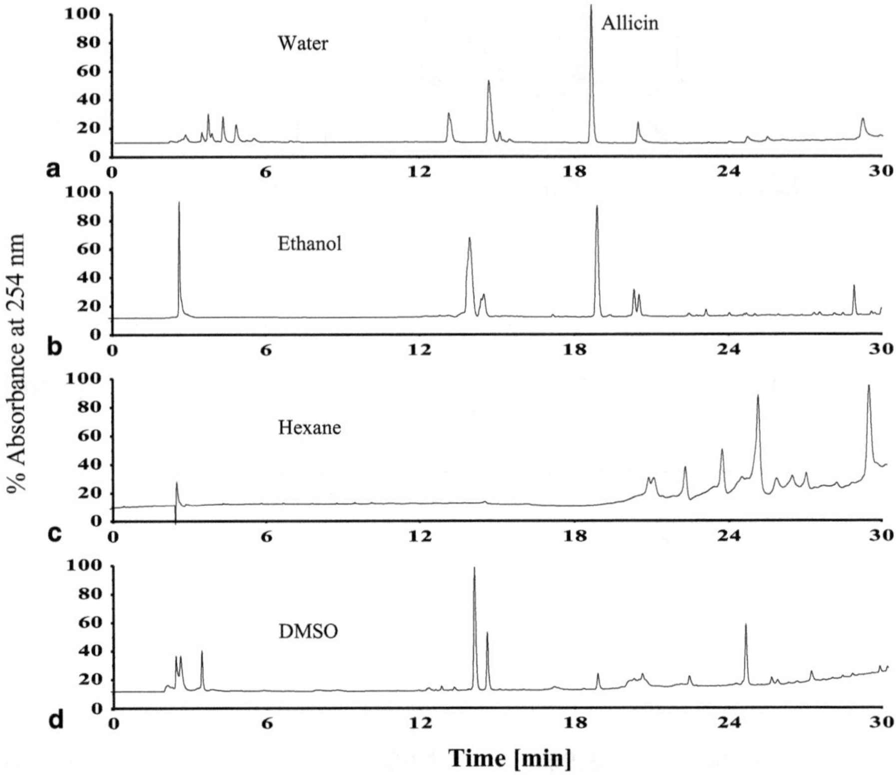

Fig. 6.6 Green RP-HPLC chromatograms of garlic extract (GE) in different solvents: (**a**) aqueous GE at pH 7. (**b**) Ethanol GE at pH 7. (**c**) Hexane GE at pH 7. (**d**) DMSO GE at pH 7. Green RP-HPLC was done under the following conditions: Instrument; Hewlett Packard 1100 (Agilent Technologies, Santa Clara, CA). Column; Kinetex RP-C_{18} (Phenomenex, Torrance, CA), 5 µm, 250 × 4,6 mm. Solvent A: 0.1 % TFA in H_2O. Solvent B: 0.1 % TFA in ethanol. Temperature: 30 °C. Flow rate: 1 mL/min. Elution: A gradient was run by holding at 5 % B for 5 min followed by 5–10 % B in 2 min, 10–35 % B in 8 min, and 35–100 % B in 15 min. Detection wavelength; 254 nm. Allicin is eluted at 18.3 min

extracts. Figure 6.6 shows that water and ethanol were the best extractants for allicin and other solvents were ineffective.

Because allicin is so unstable, it readily changes into other compounds after cooking or aging. To study the effect of cooking on allicin in garlic, we evaluated the allicin content in garlic extract after stewing or stir frying the chopped garlic in water and olive oil, respectively. Figure 6.7 shows the green RP-HPLC of different garlic extracts by boiling and stir frying the chopped garlic. The following method was used to prepare the garlic samples (following the common household cooking practice). The 23 g peeled garlic cloves were finely chopped (2×2×2 mm) using a large, sharp knife and cooked in 300 mL beaker of rapidly boiling water

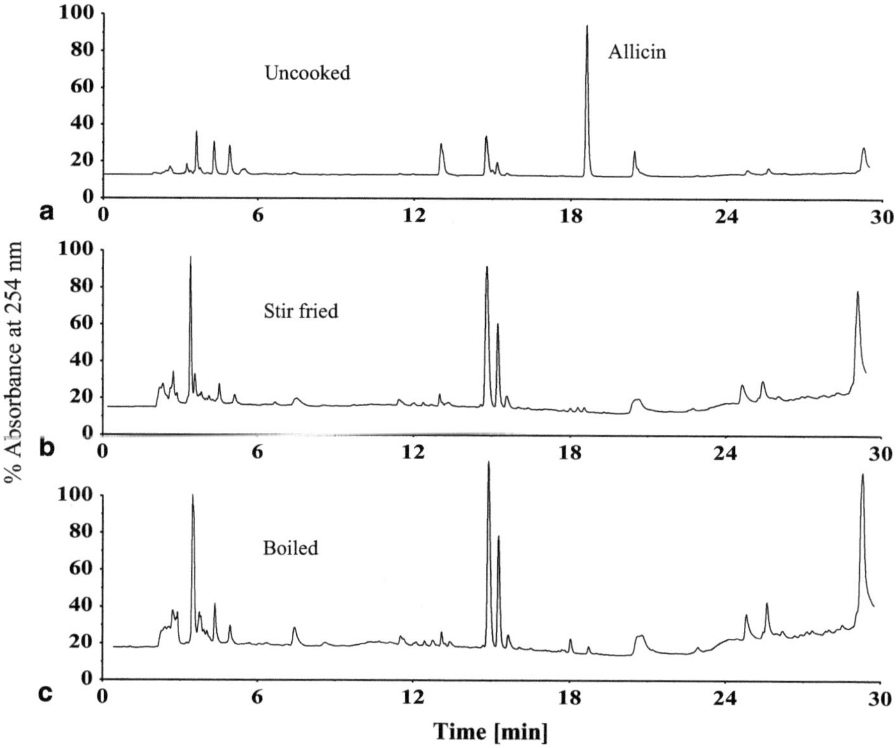

Fig. 6.7 Effect of cooking on allicin in garlic: (**a**) AGE at pH 7 without cooking. (**b**) AGE of stir fried garlic (20 min) at pH 7. (**c**) AGE of water boiled garlic (3 min) at pH 7. Green RP-HPLC was done under the following conditions: Instrument; Hewlett Packard 1100 (Agilent Technologies, Santa Clara, CA). Column; Kinetex RP-C_{18} (Phenomenex, Torrance, CA), 5 µm, 250 × 4.6 mm. Solvent A: 0.1 % TFA in H_2O. Solvent B: 0.1 % TFA in ethanol. Temperature: 30 °C. Flow rate: 1 mL/min. Elution: A gradient was run by holding at 5 % B for 5 min followed by 5–10 % B in 2 min, 10–35 % B in 8 min, and 35–100 % B in 15 min. Detection wavelength; 254 nm. Allicin is eluted at 18.3 min

(50–80 mL) for 30 min to 1 h. The solution was cooled to room temperature and the volume was adjusted to 40 mL. The resulting heated garlic sample was crushed in a blender, and the extract was prepared according to the above procedure. The stir fried sample was prepared as follows. Olive oil (3 mL) was heated in a 300 mL beaker and 23 g finely chopped garlic was stir fried for 20 min. The resulting heated garlic solution was crushed in a blender and the extract was prepared according to the above procedure. Data showed that there was very little allicin left in boiled and sautéd garlic and the anti-cancer activity is completely gone. Green HPLC is as effective as the traditional HPLC in separating and detecting the allicin. By using HPLC to purify allicin directly without organic solvent extraction, toxic waste can be eliminated.

6.5 Allicin Green Chemical Synthesis

Direct purification of allicin from aqueous garlic extract produces low overall yield. Many schemes have been proposed to produce allicin through chemical synthesis with better yields (Ilic et al. 2010; Lee et al. 2013; Cremlyn 1996; Cruz-Villalon 2001). Synthesis can start with cysteine, allyl cysteine, or diallyl disulfide (DADS). We chose to use the allyl disulfide scheme to synthesize allicin, which involves only one oxidation step and is more environmental friendly. The following procedure was used to synthesize the allicin from DADS. The synthesis of allicin is based on the oxidation of DADS with hydrogen peroxide in acidic media (Ilic et al. 2010; Lee et al. 2013; Cremlyn 1996; Cruz-Villalon 2001). All chemicals were purchased from Sigma-Aldrich Chemical Company (Saint Louis, MO) and used without further purification. Reactants of 100 μL of DADS, 126 μL 30% H_2O_2, and 200 μL acetic acid were mixed in an eppendorf tube. The tube was allowed to rotate for 10 h using a rotisserie (Barnstead International, Dubuque, Iowa) at room temperature. The solubilization step of the DADS in dichloromathine before adding H_2O_2 and acetic acid is thus eliminated in this process (Cremlyn 1996; Cruz-Villalon 2001). The DADS layer disappears completely after ~10 h as it converts to allicin. By using green RP-HPLC to purify the allicin directly from the reaction solution, we eliminate neutralization (using potassium hydroxide or $NaHCO_3$) and liquid/liquid extraction steps(using diethyl ether or dichroromethane). This leads to a greener reaction scheme which reduces the use of toxic organic solvents and improves overall yields. An additional bonus of using green RP-HPLC is that allicin is stable in the HPLC solution and can be used in further studies. Attempts to vacuum dry the pure allicin HPLC fraction failed due to the decomposition of allicin while lyophilizing. Figure 6.8 shows the time course of the green HPLC of the DADS chemical synthesis. Green HPLC is as effective as the traditional HPLC in purifying the allicin. Data clearly show that the procedure is simple, robust, effective, and eco-friendly.

6.6 Stability of Allicin

It is reported that heating or digesting the garlic reduces the effectiveness of health benefits of garlic due to the deactivation of alliinase and/or the active components (Song and Milner 2001; Cavagnaro et al. 2007; Song and Milner 1999; Lawson and Wang 2001). Allicin is generally not synthesized from garlic *in vivo* due to the irreversible deactivation of allinase below pH 3 (Freeman and Kodera 1995; Lawson and Hughes 1992; Shashikanth et al. 1985), an acidic environment typically found in the stomach. Allicin extracted from fresh garlic gradually degrade sand converts into other sulfur containing compounds within few months at room temperature. Pure allicin is expensive due to its instability. Furthermore, allicin is reported to be unstable at higher concentrations. There are limited data in the literature on the temperature, pH, and vacuum drying stabilities of allicin with conflicting results

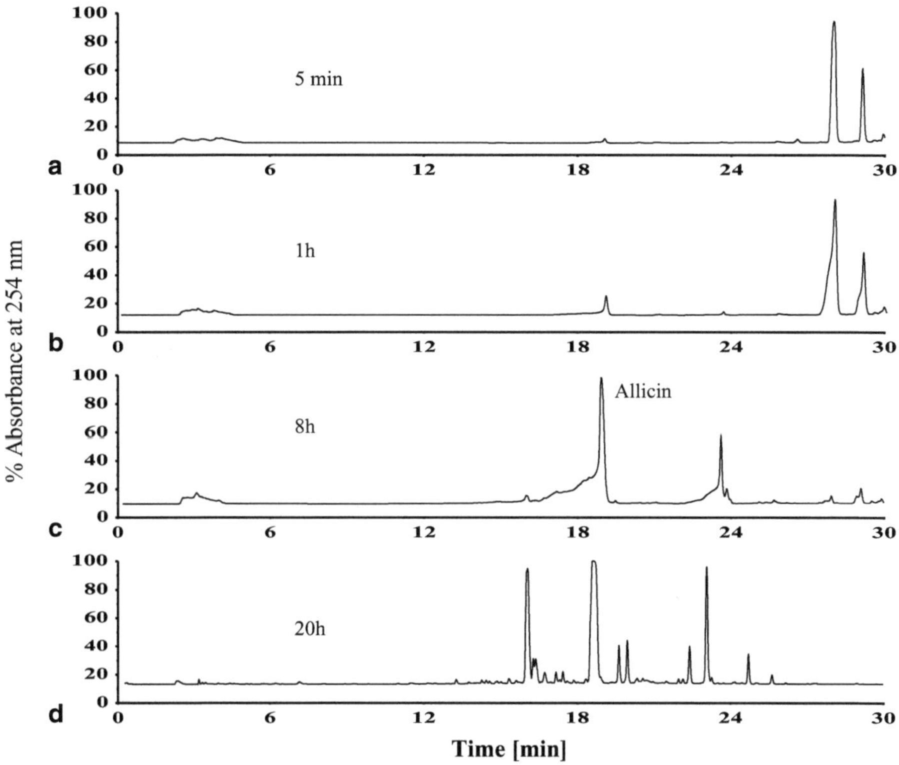

Fig. 6.8 Time course of DADS oxidation reaction at 25 °C: (**a**) 5 min. (**b**) 1 h. (**c**) 8 h. (**d**) 20 h. Green RP-HPLC was done under the following conditions: Instrument; Hewlett Packard 1100 (Agilent Technologies, Santa Clara, CA). Column; Kinetex RP-C_{18} (Phenomenex, Torrance, CA), 5 μm, 250 × 4,6 mm. Solvent A: 0.1 % TFA in H_2O. Solvent B: 0.1 % TFA in ethanol. Temperature: 30 °C. Flow rate: 1 mL/min. Elution: A gradient was run by holding at 5 % B for 5 min followed by 5–10 % B in 2 min, 10–35 % B in 8 min, and 35–100 % B in 15 min. Detection wavelength; 254 nm. Allicin is eluted at 18.3 min

(Fujisawa et al. 2008; Brodnitz et al. 1971; Hahn 1996; Yu et al. 1989; Block 1985; Amagase 2006). Here we present a systematic study of effects of temperature, pH, and vacuum drying on the stability of allicin with temperature ranging from −80 to 100 °C and pH ranging from 1 to 12. Figure 6.9 shows the stability of allicin at different temperatures by green RP-HPLC. Data show that allicin is stable at temperatures <4 °C. At 25 °C or higher, the stability of allicin will decreases. Furthermore, allicin degradation dramatically increases at temperatures >50 °C. Allicin is completely degraded after 30 min in boiling water, after 20 min stir frying in olive oil, and after overnight heating at 75 °C. In addition, allicin in garlic extract or in pure state has similar stability. Since the anti-cancer activity of the garlic extract is solely due to allicin, the IC_{50}, which represents the dilution of the AGE to produce 50 % inhibition of cells *in vitro*, of the garlic extract is also an indication of the degradation of allicin. Figure 6.10 shows the time course of the IC_{50} of the AGE and pure allicin

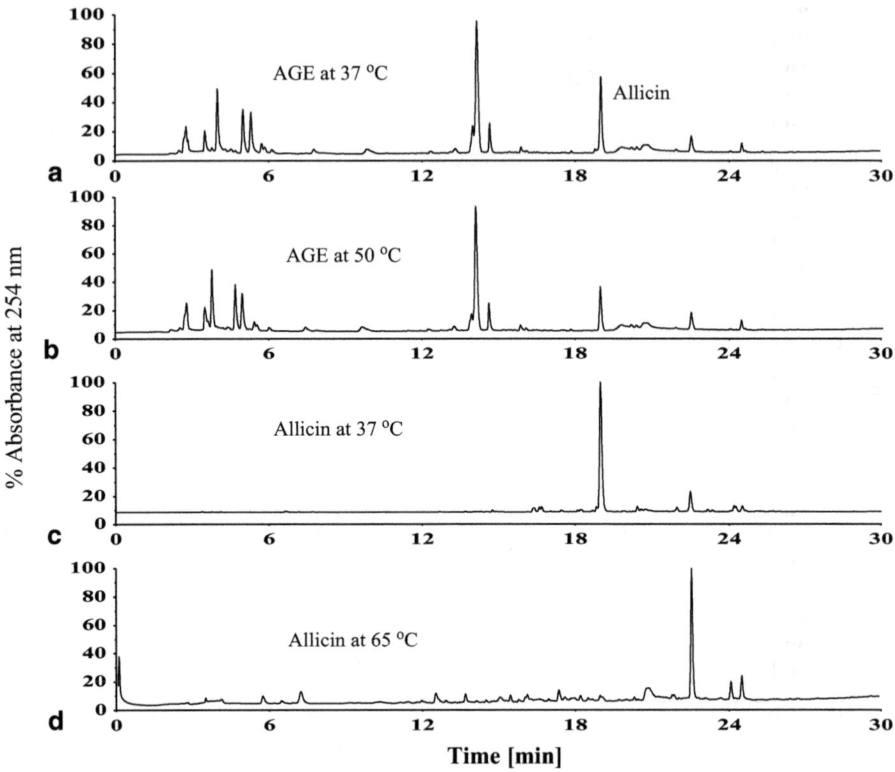

Fig. 6.9 Thermal stability of allicin: (**a**) AGE at 37 °C for 40 h. (**b**) AGE at 50 °C for 40 h. (**c**) Allicin at 37 °C for 40 h. (**d**) Allicin at 65 °C for 40 h. Green RP-HPLC was done under the following conditions: Instrument; Hewlett Packard 1100 (Agilent Technologies, Santa Clara, CA). Column; Kinetex RP-C_{18} (Phenomenex, Torrance, CA), 5 µm, 250×4,6 mm. Solvent A: 0.1 % TFA in H_2O. Solvent B: 0.1 % TFA in ethanol. Temperature: 30 °C. Flow rate: 1 mL/min. Elution: A gradient was run by holding at 5 % B for 5 min followed by 5–10 % B in 2 min, 10–35 % B in 8 min, and 35–100 % B in 15 min. Detection wavelength; 254 nm. Allicin is eluted at 18.3 min

at different temperatures ranging from −80 to 75 °C. Data show that degradation process speeds up at temperature above 50 °C, e.g., allicin is completely degraded in 20 h at 75 °C. However, allicin seems to be stable for years at temperatures below 4 °C and for months at 25 °C. The reduction in the activity of garlic extract and allicin toward *Musmusculus* colon carcinoma cells CT26.WT (ATCC, Rockville, MD) proceeded largely coincident with the quantity of allicin.

The pH is also reported to affect the stability of allicin. We tested the effect of pH on allicin ranging from 1 to 12 with green RP-HPLC. Figure 6.11 shows the stability of allicin at different pH values. Allicin is stable and forms a clear solution at pH below 7. It decomposes completely above pH 9 (opaque solution). Allicin is also not bioavailable in the blood, presumably due to the instability at blood's slightly basic pH. Figure 6.12 shows the IC_{50} of the AGE and pure allicin at different pH ranging from 1 to 12. Again, coincidence between IC_{50} and the quantity of allicin

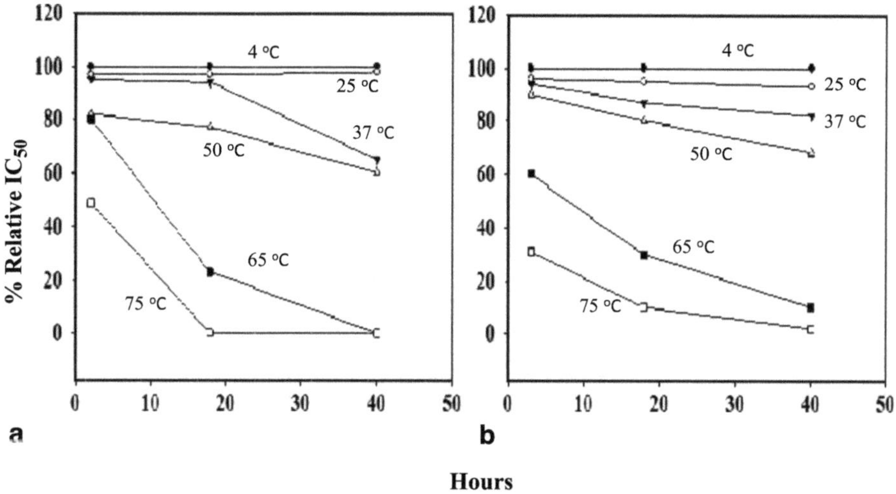

Fig. 6.10 Time course of IC_{50} on viability of *Musmusculus* colon carcinoma cells CT26.WT assessed by MTT assay(Wallert and Provost Laboratory, Minnesota State University, Moorhead, MN) of the AGE (**a**) and pure allicin (**b**) at different temperature. IC_{50} is calculated using a four parameter logistic curve fitting (SigmaPlot Software, software, Richmond, CA). The initial IC_{50} is expressed as 100%

is observed. Concentration is also shown to affect the stability of allicin. We tested the effect of vacuum drying on the stability of the allicin using green RP-HPLC. Figure 6.13 shows the green RP-HPLC of allicin solution at half dry and completely dry. Allicin seems to be stable at concentrated solution and is completely degraded after drying to a yellowish oil.

6.7 Conclusions

In the last ten years, green chemistry has gained acceptance in academia, industrial and pharmaceutical environments. The focus of green chemistry is on minimizing the harmful organic waste and maximizing the efficiency of chemical processes. This study presents effective green methodologies for preparation and purification of allicin from garlic extract and DADS acidic oxidation reaction mixture. First, the use of a green RP-HPLC using eco-friendly ethanol as the eluent. Second, the use of a simple DADS oxidation reaction scheme to eliminate the extra chemical reaction steps and toxic liquid/liquid extraction. Third, the use of a green RP-HPLC to eliminate the toxic liquid/liquid extraction step in purifying the DADS reaction mixture or AGE. Fourth, the use of a green C_{18} RP-TLC using eco-friendly 50% aqueous ethanol as the solvent. When the proposed green RP-HPLC was applied to evaluate GE sample analysis, precise results comparable to those obtained by the classical chromatographic method that uses acetonitrile as mobile phase were obtained. In

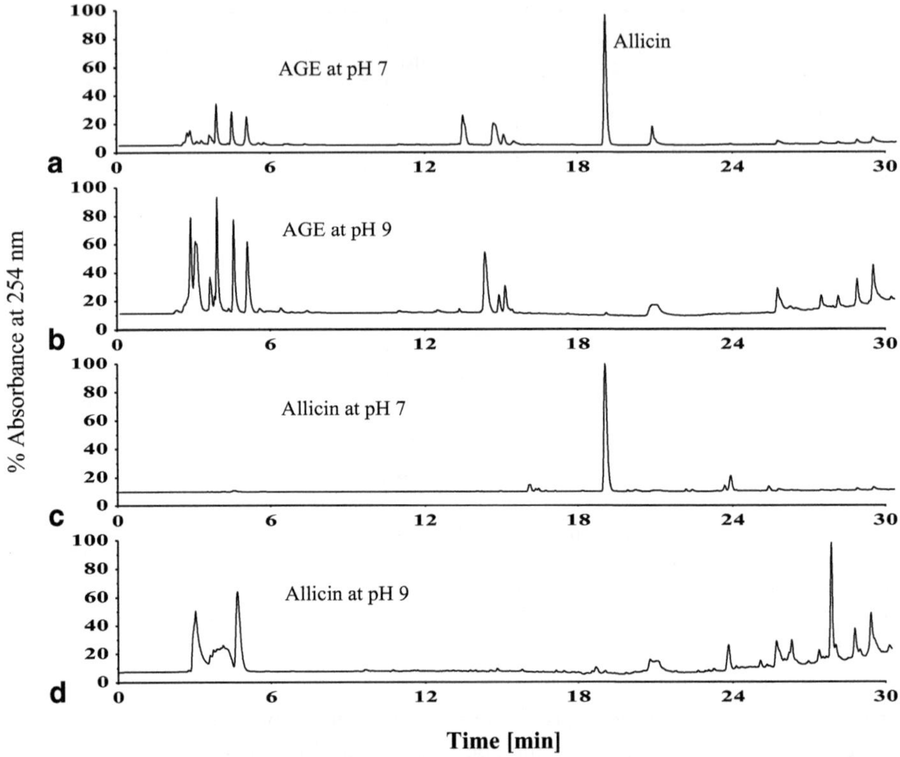

Fig. 6.11 Stability of allicin at different pH: (**a**) AGE at pH 7. (**b**) AGE at pH 9. (**c**) Allicin at pH 7. (**d**) Allicin at pH 9. Green RP-HPLC was done under the following conditions: Instrument; Hewlett Packard 1100 (Agilent Technologies, Santa Clara, CA). Column; Kinetex RP-C_{18} (Phenomenex, Torrance, CA), 5 µm, 250 × 4.6 mm. Solvent A: 0.1 % TFA in H_2O. Solvent B: 0.1 % TFA in ethanol. Temperature: 30 °C. Flow rate: 1 mL/min. Elution: A gradient was run by holding at 5 % B for 5 min followed by 5–10 % B in 2 min, 10–35 % B in 8 min, and 35–100 % B in 15 min. Detection wavelength; 254 nm. Allicin is eluted at 18.3 min

Fig. 6.12 IC_{50} of the AGE (×) and pure allicin (•) at different pH. IC_{50} is calculated using a four parameter logistic curve fitting (SigmaPlot Software, software, Richmond, CA). The IC_{50} at pH 7 is expressed as 100 %

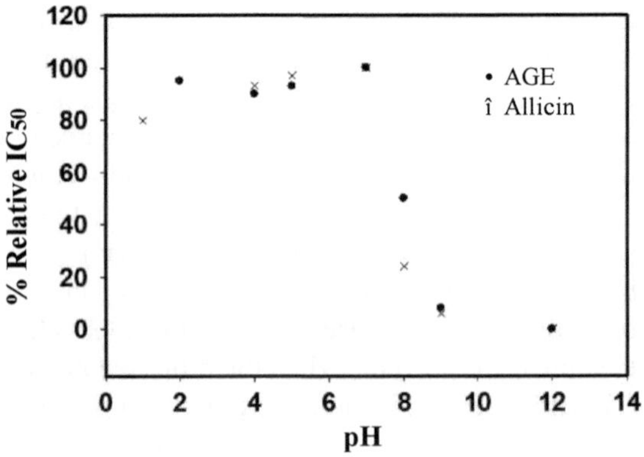

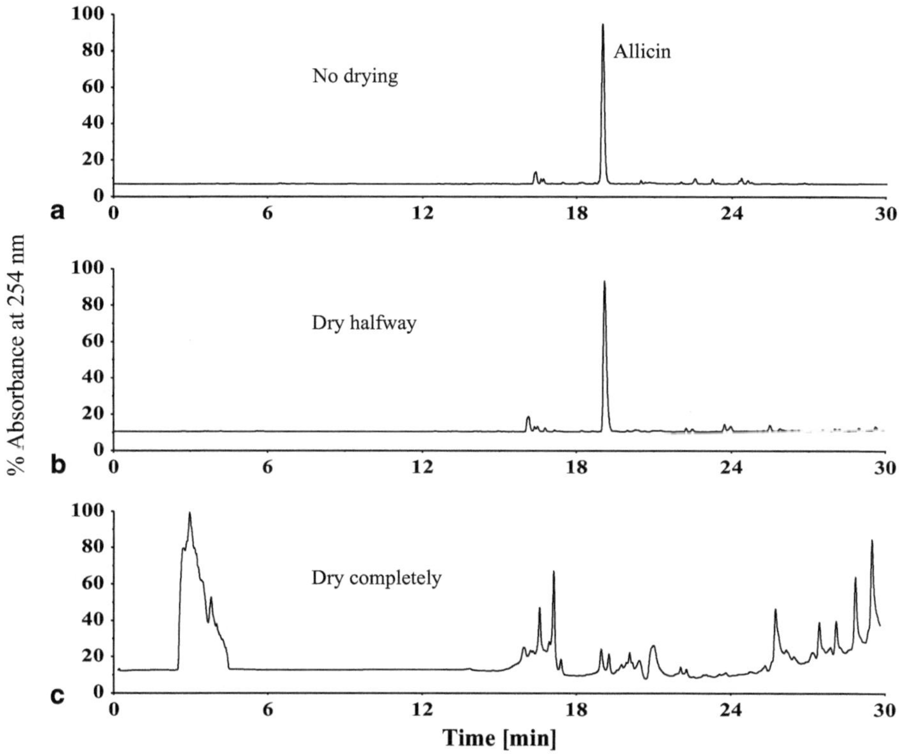

Fig. 6.13 Effect of vacuum drying on the stability of allicin: (**a**) pure allicin HPLC fraction. (**b**) Allicin HPLC fraction after the loss of half the solvent. (**c**) Yellowish allicin oil without solvent. Green RP-HPLC was done under the following conditions: Instrument; Hewlett Packard 1100 (Agilent Technologies, Santa Clara, CA). Column; Kinetex RP-C_{18} (Phenomenex, Torrance, CA), 5 μm, 250 × 4.6 mm. Solvent A: 0.1 % TFA in H_2O. Solvent B: 0.1 % TFA in ethanol. Temperature: 30 °C. Flow rate: 1 mL/min. Elution: A gradient was run by holding at 5 % B for 5 min followed by 5 to 10 % B in 2 min, 10 to 35 % B in 8 min, and 35 to 100 % B in 15 min. Detection wavelength; 254 nm. Allicin is eluted at 18.3 min

addition, the green RP-HPLC has been used to evaluate the thermal, pH, vacuum drying stabilities of allicin successfully.

Acknowledgments The authors gratefully acknowledge the support of the Research Resources Center at the University of Illinois, Chicago.

References

Ali M, Thomson M, Afzal M (2000) Garlic and onions: their effect oneicosanoid metabolism and its clinical relevance. Prostaglandins Leukot Essent Fat Acids 62:55–73

Ali M, Al-Qattan KK, Al-Enezi F, Khanafer RM, Mustafa T (2000) Effect of allicin from garlic powder on serum lipids and blood pressure in rats fed with a high cholesterol diet. Prostaglandins Leukot Essent Fat Acids 62:253–259

Amagase H (2006) Clarifying the real bioactive constituents of garlic. J Nutr 136:716S–725S
Ankri S, Mirelman D (1999) Antimicrobial properties of allicin from garlic. Microbes Infect 2:125–129
Ariga T, Seki T (2006) Antithrombotic and anticancer effects of garlic-derived sulfur compounds: a review. Biofactors 26:93–103
Block E (1985) The chemistry of garlic and onions. Sci Am 252:114–119
Block E (2010) Garlic and other alliums: the lore and the science. The Royal Society of Chemistry, Cambridge
Brettschneider F, Jankowski V, Günthner T, Salem S, Nierhaus M, SchulzA, Zidek W; Jankowski J (2010) Replacement of acetonitrile by ethanol as solvent in reversed phase chromatography of biomolecules. J Chromatogr B 878:763–768
Brodnitz MH, Pascale JV, Derslice LV (1971) Flavor components of garlic extract. J Agr Food Chem 19:273–275
Brown PR, Grushka E (eds) (2001) Advances in Chromatography 41. Marcel Dekker, New York
Cavagnaro PF, Camargo A, Galmarini CR, Simon PW (2007) Effect of cooking on garlic (Allium sativum L.) antiplatelet activity and thiosulfinates content. J Agric Food Chem 55:1280–1288
Cremlyn RJW (1996) An introduction to organosulfur chemistry. Wiley, New York
Crews P, Rodriguez J, Jaspars M (2009) Organic structure analysis. Oxford University Press, New York
Cruz-Villalon G (2001) Synthesis of allicin and purification by solid-phase extraction. Anal Biochem 290:376–378
Cutler RR, Wilson P (2004) Antibacterial activity of a new, stable, aqueous extract of allicin againstmethicillan-resistant Staphylococcus aureus. Br J Biomed Sci 61:71–74
Feng Y, Zhu X, Wang Q, Jiang Y, Shang H, Cui L, Cao Y (2012) Allicin enhances host pro-inflammatory immune responses and protects against acute murine malaria infection. Malar J 11:268
Freeman F, Kodera Y (1995) Garlic chemistry: stability of s-(2-propenyl)-2-propene-1-sulfinothioate (allicin) in blood, solvents, and simulated physiological fluids. J Agric Food Chem 43:2332–2338
Fujisawa H, Suma K, Origuchi K, Kumagai H, Seki T, Ariga T (2008) Biological and chemical stability of garlic-derived allicin. J Agric Food Chem 56:4229–4235
Gonen A, Harats D, Rabinkov A, Miron T, Mirelman D, Wilchek M, Weiner L, Ulman E, Levkovitz H, Ben-Shushan D, Shaish A (2005) Theantiatherogenic effect of allicin: possible mode of action. Pathobiology 72:325–34
Guardia M, Garrigues S (eds) (2012) Handbook of green analytical chemistry. Wiley, Chichester
Hahn G (1996) History, folk medicine, and legendary uses of garlic. In: Koch HP, Lawson LD (eds) Garlic: the science and therapeuticapplication of allium sativum L and related species (2nd ed.). Williams and Wilkins, Baltimore
Ichikawa M, Ide N, Yoshida J, Yamaguchi H, Ono K (2006) Determination of seven organosulfur compounds in garlic by high-performance liquid chromatography. J Agric Food Chem 54:1535–1540
Ilic DP, Nikolic VD, Nikolic LB, Stankovic MZ, Stanojevic LP (2010) Thermal degradation, antioxidant and antimicrobial activity of the synthesized allicin and allicin incorporated in gel. Hemijska Industrija 64:85–91
Islam MS, Kusumoto Y, Al-Mamun MA (2011) Cytotoxicity and cancer (HeLa) cell killing efficacy of aqueous garlic (Allium sativum) extract. J Sci Res 3:375–382
Kim HJ, Han MH, Kim GY, Choi YW, Choi YH (2012) Hexane extracts of garlic cloves induce apoptosis through the generation of reactive oxygen species in Hep3B human hepatocarcinoma cells. Oncol Reports 28:1757–1763
Lanzotti V (2006) The analysis of onion and garlic. J Chromatogr A 1112:3–22
Lawson LD, Hughes BG (1992) Characterization of the formation of allicin and other thiosulfinates from garlic. Planta Med 58:345–350
Lawson LD, Wang ZJ (2001) Low allicin release from garlic supplements: a major problem due to the sensitivities of alliinase activity. J Agric Food Chem 49:2592–2599

Lee J, Gupta S, Huang J, Jayathilaka LP, Lee B (2013) HPLC-MTT assay: anti-cancer activity of aqueous garlic extract is from allicin. Anal Biochem 436:187–189

Macpherson LJ, Geierstanger BH, Viswanath V, Bandell M, Eid SR, Hwang SW, Patapoutian A (2005) The pungency of garlic: activation of TRPA1 and TRPV1 in response to allicin. Curr Biol 15:929–934

Miron T, Bercovici T, Rabinkov A, Wilchek M, Mirelman D (2004) [^{3}H] Allicin: preparation and applications. Anal Biochem 331:364–369

Oommen S, Anto RJ, Srinivas G, Karunagaran D (2004) Allicin (from garlic) induces caspase-mediated apoptosis in cancer cells. Eur J Pharmacol 485:97–103

Shadkchan Y, Shemesh E, Mirelman D, Miron T, Rabinkov A, Wilchek M, Osherov N (2004) Efficacy of allicin, the reactive molecule of garlic, in inhibiting Aspergillus spp. in vitro, and in a murine model of disseminated aspergillosis. J Antimicrob Chemother 53:832–836

Shashikanth KN, Basappa SC, Murthy V (1985) Allicin concentration in the gut of rats and its influence on the microflora. J Food Sci Technol 22:110–112

Song K, Milner JA (1999) Heating garlic inhibits its ability to suppress 7, 12-dimethylbenz(a) anthracene-induced DNA adduct formation in rat mammary tissue. J Nutr 129:657–661

Song K, Milner JA (2001) The influence of heating on the anticancer properties of garlic. J Nutr 131:1054S–1057S

Yu T, Wu C (1989) Stability of allicin in garlic juice. J Food Sci 54:977–981

Yu T, Wu C, Chen S (1989) Effects of pH adjustment and heat treatment on the stability and the formation of volatile compounds of garlic. J Agric Food Chem 37:730–734

Chapter 7
Green Sample Preparation Focusing on Organic Analytes in Complex Matrices

Vânia Gomes Zuin and Cíntia Alessandra Matiucci Pereira

Abstract This chapter will describe some of the main separation techniques, considered environmentally correct, with special emphasis on sample preparation. To illustrate aspects of green analytical chemistry, techniques for sample preparation and hyphenation capable of miniaturization will be presented, which reduce, replace or eliminate the use of hazardous organic solvents focusing on organic analytes in complex matrices, mainly based on supercritical fluid extraction, membranes, solid phase extraction, matrix solid phase dispersion as well as sorptive extraction techniques, including some examples of their uses.

7.1 Introduction

Traditional approaches to reducing the risk associated to chemical activities, as proposed by the laws and environmental restrictions, have almost exclusively involved the minimization of exposure to dangerous substances (Anastas and Warner 1998). Otherwise, the green chemistry aims to reduce or eliminate this risk through the development and implementation of chemical products and processes that offer a lower or no danger, making the control of exposure unnecessary, besides, it prevents environmental impacts of accidents. In other words, if a substance presents no significant hazard, then it does not display a considerable risk, not requiring the limitation of exposure to it. It is necessary to emphasize that the danger includes, in addition to acute and chronic toxicity, carcinogenicity, genotoxicity and mutagenicity of a substance, properties such as explosiveness, flammability and corrosiveness, direct and indirect ecological impacts or persistence in the environment (Anastas 1999).

V. G. Zuin (✉)
Universidade Federal de São Carlos (UFSCar), Rodovia Washington Luiz, Km 235,
São Carlos, SP 13565-905 Brazil
e-mail: vaniaz@ufscar.br

C. A. M. Pereira
Centro Universitário Central Paulista (UNICEP), R. Miguel Petroni 5111,
São Carlos, SP 13563-470 Brazil
e-mail: cmatiucci@gmail.com

Inamuddin, A. Mohammad (eds.), *Green Chromatographic Techniques*,
DOI 10.1007/978-94-007-7735-4_7, © Springer Science+Business Media Dordrecht 2014

Since the sixties of last century, analytical chemistry has occupied a central place with the accelerated process of industrialization occurred worldwide, particularly within the environmental movement. In many different areas, the material and social progress and environmental protection activities have been inextricably linked to advances in analytical chemistry. The relevant issues such as stratospheric ozone depletion, the determination of persistent organic pollutants (POPSs) and air pollution from SOx and NOx, to name a few examples, were only able to be identified because of the remarkable progress of analytical procedures, each increasingly sophisticated instrumental techniques (Anastas and Warner 1998; Greenwood et al. 2007; De La Guardia and Garrigues 2011).

In this context, the role of analytical chemistry should not be restricted to the identification, monitoring or control of substances in different environmental compartments, but must also have regard to the prevention of pollution to the extent that their own analysis procedures do not harm the environment, both the biotic and abiotic means. Today, for the development of new analytical procedures, operating conditions, as well as the quantity and toxicity of the reactants, products and byproducts are considered as valuable as the other parameters of analytical quality, such as selectivity, sensitivity, accuracy, precision, linearity and robustness.

Despite the growing concern of all segments of society with regard to sustainability and social and environmental responsibility—as seen by the release and appreciation of international standards related to environmental management systems from companies such as ISO 14000 (International Organization for Standardization) or EMAS (EcoManagement and Audit Scheme)—most of the conventional methods employed for the determination of contaminants in various matrices have detrimental effects, mainly due to the large quantities of toxic substances involved in multiple steps of the analytical process (Kloskowski et al. 2007). Several analytical methods require hazardous substances for preservation and preparation of sample method calibration, and determination or cleaning of equipment, which may give rise to the formation of waste greater than the toxicity of the analyte investigated. To illustrate, many reference methods recommended in Brazil and abroad for the determination of organic pollutants in samples of soil or water—such as some procedures of the Environmental Protection Agency of the United States (USEPA) as the methods 3510C and 3540C—the use of 300 mL of solvent such as dichloromethane or toluene to remove the analyte by liquid-liquid extraction (LLE) or soxhlet (http://www.epa.gov/sw-846/pdfs/3510c.pdf; http://www.epa.gov/sw-846 pdfs/3540c.pdf). It is pertinent to note that the concerns about chemical analyzes also include the analyst's security and the energetic efficiency.

The activities of an analytical laboratory are commonly considered as small scale, but there are a large number of medical laboratories, environmental monitoring and control, state or private, that analyze thousands of samples daily; consequently, these labs can be compared to the pharmaceutical and fine chemicals, whose activities have generated the largest amount of industrial waste (Koel and Kaljurand 2006).

Thus, a major goal of green analytical chemistry (GAC) is the use of analytical methodologies that generate less hazardous waste, that are safer to use and environmentally friendly (Trevisan and Poppi 2006). This aim can be achieved through the development of new analytical procedures, or simply by modification or adaptation of a conventional method to incorporate steps that require less hazardous materials or at least smaller quantities (Trevisan and Poppi 2006).

Another important aspect of (GAC) is related to the area known as the process analytical chemistry (PAC), a division that involves physical and chemical determinations, including topics such as instrumentation, sampling, sample transport, communication with controllers, project management, flow engineering and chemometrics (Trevisan and Poppi 2006). The application of PAC enables to eliminate or reduce the causes of variability in a production line—by optimizing the efficient use of energy, time, raw materials, increasing quality, productivity and competitiveness of the product—and thus, can prevent pollution.

Among the twelve principles of green chemistry proposed by Anastas and Warner (Anastas and Warner 1998), there are those that are more directly related to analytical chemistry, are the prevention of waste generation, the use of safer reagents, minimizing or eliminating derivatization, the real time analysis for pollution prevention, the employment of practices that minimize potential accidents and the use of more energy efficient systems (Keith et al. 2007). In fact, no analytical procedure is entirely benign to the environment, but there is constant quest to satisfy these principles and make a method as environmentally friendly as possible.

A proposal to evaluate the influence of a particular analytical method in the environment is based on the same concepts used to evaluate different manufacturing process, in other words, the life cycle analysis (LCA). This evaluation tool takes not only the amount of reactive energy consumed in a given analysis into account, but also includes the cost and environmental impact involved, for example, to produce its reagents and solvents and the waste generated to obtain themselves (Namiesnik 2001).

As noted by Keith et al. (2007), until recently the presentation of the degree of green analytical methods, an important parameter to be considered when selecting a procedure, was not easily found in literature. In this sense, the green chemistry institute (GCI) of the American Chemical Society (ACS) has developed criteria for identifying analytical methods that use less hazardous solvents, employing safe materials and minimize waste generation. These criteria have been applied to the database of methods for the analysis of environmental samples from the USA (National Environmental Methods Index, NEMI), a catalog available on the Internet that contains information of more than eight hundred methods of analysis including procedures recommended by various regulatory agencies to monitor water, sediment or air (http://www.nemi.gov/). The green character of the analytical methods are summarized for four terms—PBT (persistence, bioaccumulation and toxicity), hazardous, corrosivity and waste generation—which consist of the principles of green chemistry more relevant to analytical chemistry. In addition to the performance characteristics (such as threshold detection, accuracy, and cost), the green profile of a method can easily be verified by the hachured quadrants, i.e. the higher the marked area the greener is the procedure.

The literature data and observation of current trends in analysis and monitoring indicate that progress has been particularly rapid in the development of precisely those methods and techniques ensuring compliance with green analytical chemistry principles; solvent-free sample preparation techniques are of particular value.

Extraction is often one of the steps in many analytical procedures. Each technique has its own merits and the choice of extraction procedure depends on several factors, including capital cost, operation cost, simplicity of operation, type and amount of extraction medium, environmental friendliness, practically and availability of standard methods (http://www.nemi.gov/). Recycling of materials and chemicals within processes is desirable. Elimination or, at least significant reduction in the amount of organic and some inorganic solvents used in sample preparation is also a priority, as it might dramatically reduce the costs of analysis, as well as the time and the labor required to process samples. Three main approaches to achieving this priority are solventless extraction, extraction using suitable of solvent or solvent assisted extraction (Koel and Kaljurand 2006; Wille and Lambert 2007; http://www.epa.gov/sw-846/pdfs/3540c.pdf; http://www.nemi.gov/.).

These steps are usually followed by chromatographic techniques, which are routinely used to separate mixtures into individual chemical species prior to the final determination. The operations involved in sample preparation are usually tedious and time-consuming and are difficult to automate (Majors 1991). An entirely green chromatography is probably impossible. Even GC, the greenest mode of chromatography, consumes purified carrier gases, the extraction of which from the atmosphere produces a measurable carbon footprint. One way to reduce energy consumption is to miniaturize the entire analytical process. The greening of analytical chemistry via miniaturization seems to be possible in principle, and the development of greener instrumentation is only limited by the creativity of the analyst (Kaljurand and Koel 2011).

7.1.1 Trends in Green Analytical Chemistry

This chapter will describe certain important separation techniques (considered environmentally correct) developed or used at present, with emphasis on sample preparation, for a typical determination of organic compounds in a hypothetical matrix. At first, often at distant places from the laboratory, the collection of samples takes place with appropriate samplers for a given set of matrix/analyte and then, the pretreatment of the sample (for example, filtration, and addition of acids or bases) aimed at maintaining the integrity and representativeness of the same. After this step, there is the sample preparation, which may include extraction (as Soxhlet, LLE, solid phase extraction, SFE), purification of the extract (as gel permeation), or analyte derivatization (http://www.epa.gov/sw-846/pdfs/3540c.pdf; http://www.nemi.gov/). In general, it is estimated that only the sample preparation consume about two thirds of the total time of analysis and is responsible for 30 % of analytical errors (Majors 1991).

Following the analyte separation and interferents from the matrix, which can occur by techniques such as chromatography (GC, HPLC) or electrophoretic (e.g., capillary zone electrophoresis), the compounds of interest can be determined using electrochemical (amperometric, coloumetric), spectroscopic (absorbance in the ultraviolet, visible, UV-VIS, fluorescence), or spectrometric (mass spectrometry, MS) methods (Kaljurand and Koel 2011).

All the steps of a method need to be revised to eliminate or reduce its impact on the environment. In this way, the separation methods may be compromised to some extent to make them green procedure. For example, liquid chromatography, one of the oldest separation techniques and used widely today on analytical, semipreparative or preparative scale, can generate a considerable amount of waste liquid (mixture containing solvents such as acetonitrile, methanol, tetrahydrofuran), especially in the latter case. Today, preparative chromatography with chiral stationary phase is considered as the more general and efficient for obtaining the enantiomers of high optical purity, i.e. a valuable tool in pharmaceutical research in the early stages of development or to the production of new drugs. The (-)-(S)-omeprazole (Nexium™) enantiomer has shown to be more efficient when used to reduce gastric secretion of gastric and duodenal ulcers than the racemic mixture. So, new alternative routes proposed by pharmaceutical companies, such as enantioselective synthesis have been revised and originated aiming to offer lower economic and environmental costs (Olbe et al. 2003).

Also, the analytical determination of chiral compounds such as (-)-(S)-omeprazole in various matrices (water, pharmaceutical products, plasma) also has required the development of new procedures which employ highly efficient separation techniques with versatile, low or practically null consumption of organic solvents, such as capillary electrophoresis (Nevado et al. 2005).

The establishment and use of miniaturized and automated instrumental methods, in order to replace the wet chemical analysis, which in the past involved primarily qualitative and semi- quantitative analyses, constitutes one of the most relevant trends in analytical chemistry (Koel and Kaljurand 2006). Along with the miniaturization, automation, eliminating or reducing the use of hazardous materials, there is also a strong trend to hyphenation of analytical techniques (coupling techniques of extraction, concentration, separation and detection) in order to obtain not only the quickest, selective and sensitive methods, but also the environmentally friendly (Koel and Kaljurand 2006). It is noteworthy that the increasing amount of data obtained with the use of instrumentation and technologies for high throughput screening has driven the development of tools for data processing, with special attention to chemometrics methods (Wilson and Brinkman 2007).

A variety of hyphenated systems for various analytical purposes can be found in the literature (Hyotylainen and Riekkola 2005). A fully automated method for the determination of triclosan and its derivative methyl triclosan in sewage water samples was developed by González-Marino et al. (Gonzalez-Marino et al. 2011). The analytes were enriched by MEPS (microextraction by packed sorbent) coupled at-line to large volume injection-gas chromatography-mass spectrometry (LVI-GC-MS). The method was accurate (RSD $<7.1\,\%$) and allowed limits of detection from 0.02 to 0.59 ng mL^{-1} in the most complex sample.

The methods for the field analytical chemistry or that allow the determination of multiple analytes in a single assay, or even dispensing the sample preparation are clearly attractive procedures. However, although a ideal green method should avoid the steps of cleaning and pre concentration of the sample but in practice the need to perform measurements in diluted samples, often below the limit of detection of an analyte or highly complex matrices (such as biological fluids, petrochemical or plants samples), which contain large amounts and variety of interfering demands the sample preparation step. In order, to illustrate aspects of green analytical chemistry, techniques for sample preparation and hyphenation capable of miniaturization will be presented, reducing, replacing or eliminating the use of hazardous organic solvents.

7.1.2 Green Techniques for Sample Preparation

The ideal situation would be the complete elimination of the sample preparation step from the analytical process. However, despite the current degree of development of the analytical instrumentation used for final determination in most instances this is not feasible. Concepts like miniaturization, integration and simplification became key concepts that have already been proved to effectively contribute to solve some of the drawbacks of conventional sample preparation methods and that, in some studies involving size-limited samples, can probably be considered the best, if not the only, analytical alternatives (Ramos et al. 2005).

7.1.2.1 Reduction and Solvent Replacement

Several techniques that make it possible to minimize or replace the organic solvents derived from nonrenewable resources such as oil, have been developed or improved for the extraction and concentration of analytes in a fast, efficient and reproducible way. In this group accelerated solvent extraction (ASE), supercritical fluid extraction (SFE), membrane extraction, ultrasonic extraction and microwave extraction may be included. While ASE employs pressure and temperature above 200 °C to perform the extractions, the others techniques employ lower temperatures, allowing the extraction of thermolabile analytes (Keith et al. 2007).

Supercritical Fluid Extraction

A supercritical fluid is a condition of the physical state of a substance which is above the critical temperature and pressure, with intermediate physicochemical properties between liquid and gas. The SFE efficiency is mainly affected by the choice of solvent, pressure, temperature, filling materials (affined to the matrix) and co-solvents. The most widely used fluid in SFE is CO_2 in the supercritical state.

At room temperature, is not toxic gas toxic at low concentrations, leaves no residue in the extract, has low cost compared to other fluids and is not flammable. It also has moderate critical conditions ($T_c = 31.1\,°C$ and $P_C = 74.8$ atm), which do not degrade thermally unstable analytes thermally unstable and facilitates the operation of the SFE system. Besides these advantages, the salvation power is high and can be higher with the introduction of polar substances to the fluid, known as co-solvents or modifiers, as methanol, water, acetone among others (Ramos et al. 2005).

An alternative to supercritical CO_2 is the extraction with superheated water. In addition to employing a solvent which is non-toxic, nonflammable and low cost, this technique offers the advantage of polarity adjustment, since the dielectric constant of pressurized water decreases dramatically with increasing temperature (100–373 °C), but high temperatures also restrict the use of the technique to some thermostable analytes, such as polychlorinated and organochlorine pesticides (Kronholm et al. 2007).

Currently, there are numerous applications of SFE in different fields, including food industries, pharmaceuticals and petrochemicals, for example, the decaffeination of coffee, tea and cocoa, the extraction of flavours and other active substances of plants (rosemary, thyme), extraction of natural dyes (β-carotene), and alkaloids (nicotine and nitrosamines in tobacco) (Hylton and Mitra 2007; Jakubowska et al. 2005).

The coupling of commercial systems for determining many organic compounds of different polarities such as polycyclic aromatic hydrocarbons (phenanthrene and pyrene), hormones (progesterone) and natural products (proanthocyanidins as catechin and epicatechin, hyperforin and its degradation products) in several matrices (water, plant extracts and urine) through SFE-HPLC-UV/MS has been recently reported in literature (Moeder et al. 2012; Poole 2003; Turner 2006; Tobiszewski et al. 2010). As an example, a novel hyphenated sample preparation method, SFE *in situ* derivatization on-line with HS-SPME-GC-MS has been developed for trapping the preservatives and antioxidants in cosmetics (Yang et al. 2010). This analytical method is simple, environmentally friendly, fast (25 min) and has a high sensitivity (RSD < 7.8 %). The limits of detection were 0.5–8.3 ng g^{-1}. According to the authors, the results were better than those obtained by using only SPME or SFE for trace preservatives and antioxidants analysis in cosmetic matrices.

Membranes

Membranes which act as selective barriers between donors and acceptor phases; are green alternative for the isolation and pre-concentration of analytes. There are two main techniques with membrane are filtration and extraction (Hylton and Mitra 2007; Koel and Kaljurand 2006). Filtration through a membrane to separate the components of a solution occurs over a pressure gradient in pore size based on the material, while the extraction with membrane mainly exploits the concentration gradients, using nonporous materials (Jakubowska et al. 2005; Koel and Kaljurand 2006).

In 2001, new adaptations of membranes include the application of microdialysis to environmental samples (Cunha et al. 2006) and the use of polypropylene devices containing solvents inside, such as cyclohexane (do Nascimento et al. 2004). In the last case, the polypropylene membranes are placed in liquid complex matrices for extraction and online determination of several organic contaminants such as organochlorine and organophosphorus pesticides, triazines, chlorophenols, polycyclic aromatic hydrocarbons (Gonzalez-Ruiz et al. 2011; Hsu and Ding 2009).

Moeder et al. (2012) developed and validated a method based on membrane-assisted solvent extraction and liquid chromatography-tandem mass spectrometry that allows the determination of 18 pesticides in red wine at minimum labour effort for sample preparation. The performance emphasizing the feasibility of the method once achieves the requirements for a sensitive and reliable method which allows the determination of the target pesticides at ngL^{-1} levels.

7.1.2.2 Solvent Elimination

A number of techniques, as SFE, enable the solvent elimination in sample preparation, once the analytes can be extracted directly from liquid or gaseous matrix for sorbent phase (http://www.epa.gov/sw-846/pdfs/3540c.pdf.). Herein the solid phase extraction (SPE), matrix solid-phase dispersion (MSPD), solid phase microextraction (SPME) and Stir bar sorptive extraction (SBSE) will be discussed.

Solid Phase Extraction (SPE)

Solid-phase extraction (SPE) is a widely used sample preparation technique for the isolation of selected analytes, usually from a mobile phase (gas, fluid or liquid). The analytes are transferred to the solid phase where they are retained for the duration of the sampling process. The solid phase is then isolated from the sample and the analytes recovered by elution using a liquid or fluid, or by thermal desorption into the gas phase. The principal goals of SPE are trace enrichment, sample clean up and transfer from the sample matrix to a different solvent or to the gas phase (Poole 2003; Turner 2006).

The trapping process is governed by the following mechanisms: adsorption of the analytes dissolved in water, on the surface of the sorbent (solid phase), mainly due to Van der Waals forces, hydrogen bonds, hydrophobic effects, π-electron interactions, and cation- and anion-exchange processes; possible dissolution of an analyte in some organic solid sorbents; and partition of analytes between the sample and liquid sorbent (liquid stationary phase), in the form of thin film, coated on or chemically bonded to a solid support (Tobiszewski et al. 2010).

SPE was initially developed as a complement or replacement for liquid-liquid extraction (LLE). Various sampling formats and sorbents have been developed to facilitate the convenient processing of different sample types and to expand the scope of the method (Kloskowski et al. 2007). A high level of automation is also possible using robotics or on-line interfaces to separation and spectroscopic instruments (Poole 2003).

As regards to liquid samples, it is probably true to say that SPE (in all of its modes and formats) is the preparation method of first choice, especially when more polar analytes are involved. An online (and automated) solid-phase extraction-liquid chromatography (SPE-HPLC) is a fully mature approach and is, moreover, the technique that can easily be miniaturized.

SPE causes the elimination or reduction of the amount of solvents used in an analytical procedure. The remarks are financial savings due to reduced purchases of high purity solvents, no need for organizing a system of collection of used solvents. Also relevant are pro-ecological factors like reduced risk of accidental disposal of used solvents to the wastewater systems, reduced risk of exposure of laboratory personnel to the vapours of volatile organic compounds; the reduction of labor and energy consumption, per analysis or per analyte (like SPE-GC, SPE-HPLC or SFE-SFC) (Namiesnik 2001). These factors have been tested for various groups of compounds (Chen et al. 2010; Cheng et al. 2011; Cunha et al. 2006; do Nascimento et al. 2004; Gonzalez-Ruiz et al. 2011; Hsu and Ding 2009; Insa et al. 2005; Papadopoulos et al. 2009; Qi et al. 2008).

The heart of SPE is the selective sorptive phase. While siloxane based chromatography phases are well established and widely used, work is still underway in characterizing these phases. Sample characteristics can also play a crucial role in the success of SPE (Raynie 2004).

SPE techniques can be applied in several modes. Typical cartridge devices consist of short columns containing a sorbent of nominal particle size of 50–60 μm, packed between porous plastic or metal frits. Low-volume cartridge or pre-column devices are the basis of on-line hyphenated systems (SPE-LC, SPE-GC) (Poole 2003).

In extraction columns the process of isolation and enrichment of analytes is realized during the flow of a liquid sample (due to gravitation or applied pressure) through the column packed with a sorbent (most often modified silica gel). In the case of samples containing suspensions, mechanical clogging of the pores of the adsorbing substance is relatively frequent, and the extraction time is rather long (Poole 2003).

The use of cartridge based SPE is well established. The same sorption chemistry is applied in new configurations, on fibers and stir bars, inside capillary tubes, on membranes, and elsewhere. The selective nature of the sorbent drives the extraction and new phases especially molecularly imprinted polymers and other types of restricted access media, are being developed (Raynie 2004).

Innovative approaches are constantly being developed. Some good ideas are the development of new SPE adsorbent materials (Covaci et al. 2010; Barker 2007); the reutilization of C18 from SPE cartridges as sorbent promoting the decrease of costs and residues; and the centrifugation in both the drying and elution steps reducing significantly the quantity of elution solvent (Kouzayha et al. 2011). Another trend is the use of miniaturized solid phase extraction (μ-SPE), once typically requires remarkably little sample material (Calvano et al. 2009, 2010; Kanimozhi et al. 2011; Yu et al. 2005).

SPE disks were developed initially to provide higher sample processing rates for large sample volumes and to minimize plugging by suspended volumes and matrix components (Poole 2003).

Empore disks (ED) are available with a diameter (47 mm) and size similar to liquid chromatography solvent filters. Their main advantages over SPE cartridges include higher sampling flow-rates, without risk of channeling, and the faster mass-transfer provided by the smaller particle sizes (Tobiszewski et al. 2009). These disks have been tested for various groups of compounds, including drugs, pyrethroids and polycyclic aromatic hydrocarbons (Ponnusamy and Jen 2011; Wu et al. 2010; Zuin et al. 2004).

Since the Empore extraction disks appear as alternatives to conventional C18 cartridges, it is of interest to assess the performance of such disks for the trace-enrichment of a variety of pesticides of different chemical groups spiked at low μg/L concentration levels, using different water types.

The use of an extraction disk is particularly easy. The membrane is placed in a filtration apparatus attached to a water-aspirator vacuum source, and the water sample is filtered through the disk after it has been conditioned with 10 mL of methanol and 10 mL of organic-free acidic water. The extraction funnel and frit assembly are transferred to a second vacuum filtration flask containing a test tube. A 5–10 mL portion of the eluting organic solvent is drawn onto the membrane, with the vacuum being interrupted so as to allow it to soak the disk for several minutes. This is normally repeated with another 5 mL aliquot of eluting organic solvent. The main advantage is the increased productivity permitted by the relatively high flow-rates. In general, the time required for the isolation of the various pesticides using disks is a half of those using cartridges (30 vs. 60 min for 1 l of water). This is significant achievement, since many samples need to be analyzed during monitoring studies and much sample preparation time can be saved.

The speed disks (SD) are modified form of ED. In SD a layer of the sorbent bed is placed between two glass-fiber filters, which considerably shorten the extraction time due to the possibility of the sample passing through at a significantly greater flow rate (Poole 2003).

Matrix Solid-Phase Dispersion (MSPD)

Matrix solid-phase dispersion (MSPD) combines both extraction and clean-up steps, and allows parallel sample processing (Aguinaga et al. 2009; Barker 2007; Campillo et al. 2009). The main difference between MSPD and SPE is that the sample is dispersed throughout the column in case of MSPD and does not retained to only first few millimeters depth. There are two possibilities: (a) the analytes are retained on the column and the interfering compounds are eluted in a washing step while, afterwards, the analytes are eluted by a different solvent or (b) interfering matrix components are selectively retained on the column, and the analytes are directly eluted (Vinas et al. 2009). The MSPD method is straightforward, rapid, and inexpensive (Navarro et al. 2002). Its attractive features are short extraction times need of small amounts of sorbent and solvent low cost, and the possibility of simultaneously performing extraction and clean up (Vinas et al. 2009).

The MSPD combined with appropriate elution solvents which ensured improved chromatographic performances, less interference in the mass spectrometer and, consequently, low LOQs. The application of MSPD is based on the blending of a semi-solid or solid sample with a solid support material (e.g., C_8/C_{18} modified silica) on its surface (Gilbert-Lopez et al. 2010; Rodrigues et al. 2010; Rubert et al. 2011). MSPD is a process for disrupting and extracting solid samples. The novelty of the technique consisted in obtaining isolation of target analytes by dispersing tissues onto a solid support, avoiding many of the difficulties encountered by employing the classical SPE approach, such as the need of sample homogenization and tissue debris removal prior to column application, as well as incomplete cell disruption (Capriotti et al. 2010). The relative polarity of the eluting solvent to that of the solid bonded-phase plays a significant role in determining what remains adsorbed on the column and what is eluted (Covaci et al. 2010).

The great interest for MSPD is due to the several advantages it offers, and its simplicity and flexibility have contributed to its diffusion over more classical sample preparation methods (Barker 2007). In contrast to classical extraction methods that require often clean-up steps, large amount of samples, sorbents and organic solvents (Kristenson et al. 2006) and thus are expensive and time consuming, MSPDE is rapid, less manual intensive, and more eco-compatible. After extraction, further sample clean-up is required or not depends on the nature of target analytes and instrumentation employed for their detection (Capriotti et al. 2010). Generally, after MSPDE, a liquid-chromatography (LC) or gas-chromatography (GC) separation is followed by mass-spectrometric determination (MS); less frequently, LC is coupled to UV or fluorescence detection, and GC to electron capture detection.

For the reasons cited above, the MSPDE protocol is considered a valid alternative to Soxhlet and MAE (Microwave Assisted Extraction), as well as SFE and PLE (Pressurized Liquid Extraction). In fact, compared to the above mentioned techniques, MSPD requires mild extraction conditions (room temperature and atmospheric pressure) providing acceptable yield and selectivity (Garcia-Lopez et al. 2008). Nevertheless, sometimes MSPD has been employed in conjunction with PLE, because the use of solvents at high temperatures and pressures can lead to increased analyte recoveries when the analytes interact strongly with the solid matrix. Moreover, MSPDE may also be improved by extracting the analytes with solvent in an ultrasonic bath before elution (Capriotti et al. 2010).

The use of inert materials for MSPD makes it a cost-effective method at the expense of selectivity which is just regulated by the molecule solubility. A considerable work performed employing inert material as dispersant, involve the use of warm-hot water as selective solvent (Bicchi et al. 2004; Duan et al. 2011).

The main advantages of this extraction procedure are the use of a non-toxic solvent, i.e., water, whose selectivity in target analyte extraction can be obtained by suitably controlling its temperature. Furthermore, only limited manipulation of the extract is required, and relatively large volumes of the final extract can be injected into a reversed phase LC column. A SPE clean-up step could eventually be added if less selective detectors than the mass spectrometer are used.

Although MSPD strategy presents many advantages, however, it is not always reported as more efficient than other techniques. It is essential to keep in mind that dealing with analysis of complex matrices, especially foods; a combination of different techniques is often needed to achieve the required performances in terms of accuracy and sensitivity (Ridgway et al. 2007).

An analytical method combining MSPD with GC/MS was successfully validated for the analysis of multi-class pesticides from a wide range of agricultural food commodities. It is a simple and rapid which can be easily miniaturized and applied to the analysis of large number of pesticides. In general, about 24 samples can be analyzed daily by one analyst. This method has been successfully applied to a set of South African fruits and vegetable samples (Bruheim et al. 2003).

Although useful for the analysis of trace contaminants in food, particularly as an aid or alternative to liquid-liquid extraction or solid phase extraction, the technique cannot be easily automated and hence it is time consuming if large number of samples are analysed. Though some MSPD extracts are clean enough for direct instrumental analysis, further clean-up step is often required, particularly with fatty matrices (Jochmann et al. 2006).

When only small amounts of the sample are available or entire extracts have to be analyzed (as in the case of online sample preparation-chromatography), extraction methods have to be miniaturized. Until now, only a few authors have used this approach for MSPD (Kristenson et al. 2006). On-line coupling of MSPD to LC or GC has been missing in the literature.

The MSPD technique with hot water as extractant has been successfully used for developing LC/mass spectrometry (MS) methods for detecting contaminants in various foodstuffs, such as animal tissues, milk, eggs, vegetables, and baby food. Furthermore, heated water provides sufficiently clean extracts and requires little manipulation (pH adjustment and filtration) before injection into a RP-LC column. In this case, large volume (0.5 mL) of the aqueous extract can be injected into the LC apparatus without peak distortion (Bicchi et al. 2004).

Some improvements such as the use of alternative sorbents, solvents, and cleanup procedures have been introduced. Efforts have also been made to develop automatic or semi-automatic MSPD procedures. Combining PLE with MSPD results in further improving the sensitivity and selectivity of MSPD-based methods by working at elevated temperature and pressures. When combining MSPD with RP-LC, the use of heated water, an inexpensive and environmentally friendly solvent, as extractant has drastically simplified the analytical procedure for analyzing target compounds in foodstuffs. When analysing aqueous foodstuff matrices, such as milk and fruit juices, an advantage offered by the MSPD technique is the use of water—miscible solvents, water (alone or mixed with polar solvents) for extracting target compounds is not precluded (Bicchi et al. 2004). In summary, MSPD is a simple, user-friendly technique that has proved to be extremely practical for a wide variety of analyte/matrix combinations (Bogialli and Di Corcia 2007).

Sorptive Extraction Techniques

Sorptive extraction techniques are based on the distribution equilibria between the sample matrix and sorptive materials. Much attention has been paid to solvent-free sample preparation techniques that are based on sorptive extraction using a polymeric gum.

In general, the SPME and SBSE techniques are coupled to GC.In former case, SPME fibers are inserted in the GC heated injectors whereas the SBSE techniques are introduced in specific systems to the analyte thermodessorption prior to the chromatographic separation. The SPME and SBSE methods have shown significant suitability for various analytical applications, as example, biomedical, environmental, natural products, food and drinks (Pico et al. 2007).

Solid Phase Microextraction (SPME)

This technique is based on partitioning of an analyte between a sample and a fused-silica fiber coated with stationary phase, which can be a liquid polymer or a solid polymer dispersed in the liquid polymer. The analyte is then desorbed from the fiber directly into a suitable separation and detection system (Tobiszewski et al. 2009). The most common applications are in the field of organic-compound determination (i.e. volatile organic compounds, alkylbenzenes, polycyclic aromatic hydrocarbons and pesticides as target analytes) (Eom et al. 2008; Alonso et al. 2011; Zygmunt et al. 2001; Ishizaki et al. 2010; Melo et al. 2011; Saito et al. 2010; Tholl et al. 2006; Bicchi et al. 2002; Barriada-Pereira et al. 2010; Iparraguirre et al. 2011; Klein et al. 2010).

SPME uses a fused silica fiber that is coated on the outside with an appropriate stationary phase (a 5–100 μm thick coating of different polymers, e.g. polydimethylsiloxane, PDMS) but today, a variety of fibers, with different thickneses and polarities are available, and good recoveries and repeatability are achieved in several applications (Ponnusamy and Jen 2011; Qi et al. 2008; Wu et al. 2010; Zuin et al. 2004). The small size of the SPME fiber and its cylindrical shape enable it to get fitted well inside the needle of a syringe-like device. Target molecules from a gaseous or a liquid sample are extracted and concentrated on the polymeric fiber coating. SPME coupled to GC and GC-MS, as well as to HPLC and LC-MS has been used (Turner 2006).

SPME can either be performed by headspace extraction (HS-SPME) by placing the fiber in the vapour above a gaseous, liquid or solid sample or by direct immersion extraction (DI-SPME), by immersing the fiber in a liquid sample. After a certain extraction time, the SPME needle is removed from the septum and inserted into the injection port of the GC or into the desoprtion chamber of the SPME-HPLC interface. The main advantage of SPME compared to LLE and solid phase extraction (SPE) are: need of little solvent easily automated technique, coupling to chromatographic techniques such as GC and HPLC, the combining of sampling and extraction into one step and the ability to examine smaller sample sizes (Aguinaga et al. 2009; Campillo et al. 2009; Ridgway et al. 2007). It can be used for both polar and non-polar analytes in a wide range of matrices by coupling GC and/or LC.

Some disadvantages of SPME include batch to batch variation and robustness of fiber coatings (Ridgway et al. 2007). The lack of sensitivity due to the extremely small amount of sorbent that can be coated onto the SPME fiber in combination with the low analyte capacity of the SPME sorbents compared to many other typical adsorbents is another major disadvantage of this technique (Turner 2006).

Several parameters such as extraction mode (HS-SPME or DI-SPME), type and thickness of fiber coating, extraction time, sample properties (analyte concentration, pH, buffer, temperature, agitation) and analyte desorption have to be considered during the application of SPME (Zygmunt et al. 2001; Qi et al. 2008).

Recently, some authors have been used the SPME in derivatization process to obtain volatile derivatives (Vinas et al. 2009, 2010). The process has been called on-fibre derivatization and consists of direct immersion SPME to adsorb the analytes and in sequence the fiber is placed in the headspace of the derivatizing reagent. This process significantly reduces the amount of solvent that would be used in derivatization.

There are several classifications for these SPME based microextraction techniques. Duan et al. (2011) classified into sample-stir microextraction (SSME) (static) and sample-flow microextraction (SFME) (dynamic) (Duan et al. 2011).

Sample-Stir Microextraction (SSME)

SSME includes fiber SPME, thin-film SPME and stir-bar sorptive extraction (SBSE). These microextraction procedures are typically carried out in stirred samples.

Because of the fragility of the silica fiber, a particular casing or shell was required to protect the coated fiber from physical collision during extraction (Duan et al. 2011). For air samples, the fiber can be fixed in a chamber containing gas samples for extraction.

Thin-film microextraction (TFME) utilizes a membrane carrying the extraction phase. Its robustness is better than that of fiber SPME. TFME has higher extraction efficiency and load ability due to the larger volume of its extraction phase and its higher surface-to-volume ratio. A typical thin film is a polymer (e.g., PDMS). Just as in SBSE, the thermal desorption of analytes adsorbed in the film needs the assistance of distinct units (e.g., a liner). In on-site application, after extraction, the film is usually transported to the laboratory for analysis. However, it is still a technique little used and applications are limited to the analysis of water spiked with standards (Bruheim et al. 2003). A recent development of TFME for on-site extraction involves a coated hollow-fiber (HF) membrane (Basheer et al. 2007). The inner and outer surfaces of the HF are coated by a thin layer (1.2–1.3 μm) of polymer that has a strong affinity with chlorinated and brominated compounds. The coated membrane can be placed directly into a sample vial containing seawater to extract analytes on site (Duan et al. 2011).

Sample-Flow Microextraction (SFME)

In SFME, dynamic sampling is performed by a syringe or pump that introduces the sample passing through the coatings or sorbents either in a needle, a tube or a tip.

Agitation of the sample is avoided for these types of SPME. SFME includes in-needle, in-tube and in-tip SPME.

In Solid Phase Dynamic Extraction (SPDE) the volume of the coating on the inner wall of the needle is several μL, compared with 0.5 μL for an SPME fiber, so SPDE achieves a higher enrichment factor. Moreover, the device is more robust than the fragile SPME fiber, since the needle is hard to damage mechanically. The main drawback of SPDE is that it tends to suffer carry-over, because the analytes tend to remain in the inner wall of the needle after thermal desorption (Mieth et al. 2009) which limits the analytical application of SPDE in the laboratory and on site. According to the literature, the number of SPDE applications is increasing year after year. Some examples are the analysis of drugs and volatiles in different matrices (Bicchi et al. 2004; Jochmann et al. 2006; Musshoff et al. 2002).

Where the sorbents or fibers are packed into a removable needle mounted on a gas-tight syringe the technique is called Packed-needle microextraction (PNME). The needle packed with sorbent particles, also known as sorbent-packed needle-trap device (SP-NTD), is used for gas samples or in the headspace of liquid samples. Its extraction capacity can be easily improved by increasing sorbent and/or sampling volumes. The SP-NTD does not need cryogenic focusing during thermo desorption, because of the low mass of sorbent in the narrow needle and fast desorption. The SP-NTD method is therefore suitable for on-site sampling and analysis. Recent developments demonstrated the capability of SP-NTD for on-site sampling of volatile compounds (Fan et al. 2011; Yamaguchi and Lee 2010). The needle device after extraction can be stored at room temperature for a week without significant sample loss (Eom et al. 2008).

According to Alonso et al. (Alonso et al. 2011) the use of the needle-trap resulted in a significant concentration factor that led to a reduction in the lowest concentration detected. The combination of needle-trap devices and headspace sampling has given improved method sensitivity, allowing LODs in the range of ngL^{-1} to be reached.

In Fiber-packed needle-trap device (FP-NTD) the sorbent particles are replaced by a bundle of polymer-coated fibers, being more useful for liquid samples. The extraction by the needle is carried out by pumping the sample solution form the syringe into the needle at a constant flow rate. The extracted analytes in the needle can be desorbed thermally or by passing a small amount of desorption solvent through the needle for injection. Since the extracted samples could be quantitatively determined after storage of the sampled needles at room temperature for at least three days, this type of SPME also has considerable potential for on-site sampling (Zygmunt et al. 2001).

The Microextraction by Packed Sorbent (MEPS) inserts the sorbent (1 mg) directly into the syringe (100–250 μL) barrel as a plug or between the needle and the barrel as a cartridge but not in the needle. This technique can be seen as miniaturized solid-phase extraction (SPE) that has a procedure similar to SPDE and NTD. The sample solution is pulled and pushed through the syringe several times. After washing to remove interfering compounds, the extracted analytes are eluted by a small volume of solvents (10–50 μL) directly into the GC or LC injector. An important advantage of MEPS is that it can handle small sample volumes (10 μL) as well as large ones (1.000 μL) (Ishizaki et al. 2010; Mieth et al. 2009), so it is very promising for application to biological and environmental samples (Duan et al. 2011). A limitation of MEPS for on-site extraction is that the sorbent must be conditioned by solvent before extraction (Duan et al. 2011).

In-tube-solid-phase microextraction using an open tubular fused-silica capillary column with an inner surface coating as the SPME device is simple and can be easily coupled on-line with HPLC and LC-MS. In-tube SPME allows convenient automation of the extraction process, which not only reduces the analysis time, but also provides better accuracy, precision, and sensitivity than manual off-line techniques (Ishizaki et al. 2010; Melo et al. 2011; Saito et al. 2010). This technique was developed due to the difficulties of interfacing SPME with LC systems. A disadvantage of in-tube devices is that particles need to be removed from samples prior to extraction (by filtration or centrifugation) (Basheer et al. 2007).

SPME has been used for the study of endogenous and exogenous volatile organic compounds in various samples, including medicinal and aromatic plants (Tholl et al. 2006). Bicchi et al. (2006) evaluated the performance of the SPME method based on the headspace in static mode for the quality control of sage (*Salvia lavandulifolia* Vahl.) and chamomile (*Matricaria chamomilla* L.) by means of coated fibers with PDMS and carboxen/divinylbenzene/PDMS (CAR/DVB/PDMS). The results showed that the SPME associated with GC flame ionization detection and mass spectrometry can be used for the routine analysis of plant tissue in which the fiber CAR/DVB/PDMS showed the best performance.

Stir-Bar Sorptive Extraction

The technique uses stir bars, usually 10 or 40 mm long and coated with polydimethylsiloxane (PDMS). For extraction, the stir bar is introduced into a vessel with the water sample being analyzed. Stirring results in the faster analytes portioning between the matrix and the stationary phase (PDMS) coated on the iron stir bar. After extraction is completed, the stir bar is removed from the water sample, and the analytes dissolved in the bar coating are released by thermal desorption or solvent extraction. This technique enables the effective extraction and hence sensitive determination of volatile and semi-volatile organic compounds in aqueous samples. When coupled with thermal desorption (TD)-GC-MS, detection, limits down to the low ng/L range can be reached. Because the phase ratio (volume of the water phase divided by the volume of the PDMS phase) is much lower than in SPME, higher recoveries and hence higher sensitivities have been obtained, especially for volatile compounds (Zygmunt et al. 2001). In SPME and SBSE, reusable elements (SPME-fiber, SBSE-stir bar) are used, and that is also important from the point of view of green chemistry (Tobiszewski et al. 2009).

Adaptations of the sorbents look to improve the extraction efficiencies of polar analytes and to allow their use in novel formats. Sorbent placements include stir bar sorption (Keith et al. 2007).

Although each individual chromatographic separation may use only a few milliliters of solvent, chromatographic separations can use an enormous amount of solvents annually. Adapting solid phases to allow water as the mobile phase is a significant step toward greening (Keith et al. 2007).

SBSE has many similarities to SPME, as it is also a solventless sample preparation technique, and it uses similar sorbents (based on PDMS). In SBSE, an aqueous sample is extracted by stirring for a certain time with a PDMS-coated stir-bar. The stir-bar is thereafter removed from the sample, and the absorbed compounds are then either thermally desorbed and analyzed by GC-MS, or desorbed by means of a liquid for interfacing to a LC system. Heat-desorption gives higher sensitivity while liquid desorption provides higher selectivity (Turner 2006).

Headspace sorptive extraction (HSSE) is a similar technique developed by Bicchi et al. (Bicchi et al. 2002). In HSSE, a PDMS stir-bar is used for headspace sampling of volatile organic molecules. This technique also has similarities to HS-SPME (Turner 2006). Most SBSE or HSSE applications involve the use of thermal desorption followed by GC to recover the analytes accumulated in the coated stir-bar, which implies not using organic solvents and allows the complete introduction of the extracted solutes in the chromatographic system (Barriada-Pereira et al. 2010; Iparraguirre et al. 2011; Klein et al. 2010; Morlock and Kopacz 2008; Ochiai et al. 2005, 2006; Ridgway et al. 2010; Sasamoto et al. 2007; Yamaguchi and Lee 2010; Zuin et al. 2006).

A stir bar in SBSE is coated with up to 125 μL PDMS, which enables quantitative extraction of many organic compounds from aqueous samples of 10–100 mL in volume. This should be compared to the situation in SPME, where the maximum volume of PDMS that can be coated onto the fiber is around 0.5 μL (100 μm thick coating). Hence, compared to SPME, SBSE offers a favorable alternative for quantitative analysis with much higher sensitivity and repeatability. Another advantage is that just as with SPME, no organic solvent is required for the extraction. However, SBSE is not as easily coupled on-line to other separation techniques such as GC and HPLC (Turner 2006).

Common applications for SBSE (and HSSE) are the analysis of PAHs in drinking water and juices (Popp et al. 2003; Sandra et al. 2003; Zuin et al. 2006), flavors and off-flavors in food samples (Fan et al. 2011; Franc et al. 2009; Guerrero et al. 2011), alkylphenols (Iparraguirre et al. 2011), food taints (Hampel et al. 2005). The technique has also widely been used for pesticide analysis in sample matrices like wine (Sandra et al. 2001), grapes (Juan-Garcia et al. 2004), honey (Blasco et al. 2004) and other food matrices (Ochiai et al. 2005, 2006; Sandra et al. 2001, 2003; Sasamoto et al. 2007).

To illustrate the applicability of the technique for real samples (Bicchi et al. 2003; Zuin et al. 2005, 2006), a method based on SBSE in combination with a system for thermo desorption (TD) coupled to a gas chromatograph and a mass spectrometer (TD-GC-MS) was used successfully for the determination of eight insect repellents widely used (such as dimethyl and N, N-diethyl-m-toluamide) in samples of river water, lake and waste water. The method allowed limits of detection of 0.5–3.0 ng/L (except dimethyl—150 ng/L), showed good linearity (0.997) and reproducibility (DPR<20%) (Rodil and Moeder 2008).

Liquid desorption (LD) is an alternative to thermal desorption when thermally labile solutes are analysed, when the separation is carried out using liquid chromatography or capillary electrophoresis. During LD mode, the polymer-coated stir-bar is immersed in a stripping solvent or solvent mixture for the chemical desorption of the extracted solutes. The minimum stripping solvent volume must guarantee the complete immersion of the coated stir-bar and, obviously, the solvents or mixtures used in this step must be compatible with the polymer. Acetonitrile and methanol are the most common desorption solvents, but mixtures with water or aqueous buffer have also been studied (Prieto et al. 2010). LD offers additional attractive features such as cost-effectiveness, the opportunity for method development and possible re-analysis (Serodio and Nogueira 2006).

Compared to LC, few reports have been proposed to combine SBSE with CE (De Villiers et al. 2004; do Rosario and Nogueira 2006; Juan-Garcia et al. 2004), although CE offers the high resolution required in the analysis of complex matrices with a reduced amount of sample, reduced solvent, reagent consumption and waste generation.

The factors, which affect the extraction and desorption in SBSE, are pH, ionic strength, temperature, stirring rate, time of extraction or desorption. These factors must be optimized and the application of Designs of Experiments (DoE) is recommended. This approach uses a series small of carefully designed experiments which allow a thorough exploration of the experimental space. The application of DoE methodology has made the identification of interaction among different variables of the SBSE extraction process possible (Prieto et al. 2010).

The experimental design methodology is the first choice to evaluate a large number of parameters and to optimize experimental conditions. It is based on the use of an optimum and reasonable set of experiments which allows varying all experimental factors simultaneously and, taking into account the possible interactions between the factors besides reducing the number of experiments, to conclude the best extraction (Bourdat-Deschamps et al. 2007).

Some novel applications of SBSE are described in the literature. SBSE as preconcentration of solid samples where the sample is not previously extracted and a direct SBSE extraction is performed (Bicchi et al. 2003; Burkhardt and Mosandl 2003; Hampel et al. 2005; Kreck et al. 2002, 2003; Sewenig et al. 2005; Tan et al. 2008). SBSE for multi-residue analysis solved by multi-shot or dual mode or in the sequential modes. In the multi-shot mode different sample aliquots are extracted under the same or different extraction condition using a coated stir-bar per sample and, then, the stir-bars are simultaneously desorbed in the TDU unit (Kawaguchi et al. 2004). When the different sample aliquots are extracted under the same chemical conditions, an increase in sensitivity is only searched for certain analytes that are best extracted under the same conditions. In sequential SBSE, the extraction conditions of a single sample aliquot are modified depending on the analytes to be extracted (Ochiai et al. 2008) using one or more stir-bars.

The major advantage of the stir bar technique is the high concentration factors that can be achieved. It can be used for liquid or semi-solid complex matrices and therefore has potential for many applications in food analysis. Zuin et al. (Zuin et al. 2006) compared SBSE to membrane assisted solvent extraction (MASE) for the determination of pesticide and benzo[a]pyrene residues in Brazilian sugarcane juice.

They found that generally faster analysis and better recoveries were achieved using MASE, whereas greater sensitivity and repeatability were obtained with SBSE.

Currently only the PDMS coating is commercially available, making the technique most suited to non-polar analytes from aqueous media. However, to improve the recovery of more polar analytes, "dual-phase twisters" which combine both absorption and adsorption, have been described by Bicchi et al. (2005) for SBSE and HSSE. These stir bars consist of short PDMS tubes closed at both ends with magnets, with an inner cavity that is packed with activated carbon adsorbent. They were shown to improve the recovery of volatile and/or polar compounds when compared to conventional PDMS stir bars in the analysis of coffee, sage and whisky.

Some strategies are found in the literature to improve the extractability of polar compounds, among them the dual mode SBSE extraction. Dual mode is performed simultaneously on two aliquots of a sample under different extraction conditions using a coated stir bar per sample. Ochiai et al. (2005) optimized and validated a dual mode SBSE extraction for the determination of pesticides in vegetables, fruit and tea. Ochiai et al. also described a dual SBSE method for the analysis of pesticides multiresidues in aqueous solution (Ochiai et al. 2006). Two extractions were performed simultaneously: one extraction with 30% NaCl, mainly targeting solutes with log $K_{o/w}<3.5$ and another, without addition of NaCl, targeting solutes with log $K_{o/w}>3.5$.

SBSE is advantageous in dealing with very dilute media and trace concentration samples; compared to SPME, SBSE generally yields better detection limits. On the other hand, SPME can be fully automated which is not yet completely true for SBSE (Sanchez-Rojas et al. 2009).

SBSE can be used as a convenient sampling system which can be easily transported for subsequent laboratory analysis or, even better, for on-site determination of the target compounds when combined with portable and miniaturized instrumentation (Ramos 2012). Although SBSE exhibits several advantages and potentials for on-site sample preparation, it led to few publications about on-site application of SBSE to real samples. The major reasons for this are the portability of the device, the time spent for extraction and the limitations of the PDMS phase, which are still currently limited to non-polar or semi-polar analytes. Recent development of SBSE is focusing on new-generation materials that can extend its range of applications (Duan et al. 2011).With advances in coating materials, on site extraction by SBSE may increase gradually in the future. Also, there are many issues that require further research in the methodology as well as in the automatisation.

7.2 Conclusions

The use of Green Chemistry principles in development of new techniques and methodologies is extremely important to reduce the environmental impacts of chemical analyses, in manufacturing activities, research or regulation, from public or private institutions. Coupled with the need to ensure analytical quality, obtaining accurate and precise results, we should also make every step of the method as environmentally friendly as possible. Besides the sample preparation with particular emphasis on those that minimize or eliminate the use of toxic organic solvents, the separation systems, particularly chromatographic and electrophoretic techniques, and data acquisition, such as spectroscopic, electrochemical and bioanalytical also offer many opportunities and potential towards the green analytical practices.

References

Anastas PT (1999) Green chemistry and the role of analytical methodology development. Crit Rev Anal Chem 29(3):167–175

Anastas PTW, Warner JC (1998) Green chemistry: theory and practice. Oxford University Press, New York

Aguinaga N, Campillo N, Vinas P, Hernandez-Cordoba M (2009) Solid-phase microextraction coupled to gas chromatography-mass spectrometry for the analysis of famoxadone in wines, fruits, and vegetables. Spectrosc Lett 42(6–7):320–326. doi:Pii 917313600, 10.1080/00387010903178723

Alonso M, Cerdan L, Godayol A, Antico E, Sanchez JM (2011) Headspace needle-trap analysis of priority volatile organic compounds from aqueous samples: Application to the analysis of natural and waste waters. J Chromatogr A 1218(45):8131–8139. doi:10.1016/j.chroma.2011.09.042

Barriada-Pereira M, Serodio P, Gonzalez-Castro MJ, Nogueira JMF (2010) Determination of organochlorine pesticides in vegetable matrices by stir bar sorptive extraction with liquid desorption and large volume injection-gas chromatography-mass spectrometry towards compliance with European Union directives. J Chromatogr A 1217(1):119–126. doi:10.1016/j.chroma.2009.10.076

Barker SA (2007) Matrix solid phase dispersion (MSPD). J Biochem Biophys Methods 70(2):151–162. doi:10.1016/j.jbbm.2006.06.005

Basheer C, Vetrichelvan M, Valiyaveettil S, Lee HK (2007) On-site polymer-coated hollow fiber membrane microextraction and gas chromatography-mass spectrometry of polychlorinated biphenyls and polybrominated diphenyl ethers. J Chromatogr A 1139(2):157–164. doi:10.1016/j.chroma.2006.11.008

Bicchi C, Iori C, Rubiolo P, Sandra P (2002) Headspace sorptive extraction (HSSE), stir bar sorptive extraction (SBSE), and solid phase microextraction (SPME) applied to the analysis of roasted arabica coffee and coffee brew. J Agr Food Chem 50(3):449–459. doi:10.1021/Jf010877x

Bicchi C, Cordero C, Iori C, Rubiolo P, Sandra P, Yariwake JH, Zuin VG (2003) SBSE-GC-ECD/FPD in the analysis of pesticide residues in Passiflora alata Dryander herbal teas. J Agric Food Chem 51(1):27–33. doi:10.1021/jf020630f

Bicchi C, Cordero C, Liberto E, Rubiolo P, Sgorbini B (2004) Automated headspace solid-phase dynamic extraction to analyse the volatile fraction of food matrices. J Chromatogr A 1024(1–2):217–226. doi:10.1016/j.choma.2003.10.009

Bicchi C, Cordero C, Liberto E, Rubiolo P, Sgorbini B, David F, Sandra P (2005) Dual-phase twisters: a new approach to headspace sorptive extraction and stir bar sorptive extraction. J Chromatogr A 1094(1–2):9–16. doi:10.1016/j.chroma.2005.07.099

Blasco C, Fernandez M, Pico Y, Font G (2004) Comparison of solid-phase microextraction and stir bar sorptive extraction for determining six organophosphorus insecticides in honey by liquid chromatography-mass spectrometry. J Chromatogr A 1030(1–2):77–85. doi:10.1016/j.chroma.2003.11.037

Bogialli S, Di Corcia A (2007) Matrix solid-phase dispersion as a valuable tool for extracting contaminants from foodstuffs. J Biochem Biophys Methods 70(2):163–179. doi:10.1016/j.jbbm.2006.07.007

Bourdat-Deschamps M, Daudin JJ, Barriuso E (2007) An experimental design approach to optimise the determination of polycyclic aromatic hydrocarbons from rainfall water using stir bar sorptive extraction and high performance liquid chromatography-fluorescence detection. J Chromatogr A 1167(2):143–153. doi:10.1016/j.chroma.2007.08.025

Bruheim I, Lord H, Pawliszyn J (2003) Comments on "Helical sorbent for fast sorption and desorption in solid-phase microextraction-gas chromatographic analysis". Anal Chem 75(15):3946–3949. doi:10.1021/Ac0264271

Burkhardt D, Mosandl A (2003) Biogenetic studies in Syringa vulgaris L.: bioconversion of O-18(H-2)-labeled precursors into lilac aldehydes and lilac alcohols. J Agric Food Chem 51(25):7391–7395. doi:10.1021/Jf0304674

Calvano CD, Jensen ON, Zambonin CG (2009) Selective extraction of phospholipids from dairy products by micro-solid phase extraction based on titanium dioxide microcolumns followed by MALDI-TOF-MS analysis. Anal Bioanal Chem 394(5):1453–1461. doi:10.1007/s00216-009-2812-y

Calvano CD, Aresta A, Zambonin CG (2010) Detection of hazelnut oil in extra-virgin olive oil by analysis of polar components by micro-solid phase extraction based on hydrophilic liquid chromatography and MALDI-ToF mass spectrometry. J Mass Spectrom 45(9):981–988. doi:10.1002/Jms.1753

Campillo N, Penalver R, Lopez-Garcia I, Hernandez-Cordoba M (2009) Headspace solid-phase microextraction for the determination of volatile organic sulphur and selenium compounds in beers, wines and spirits using gas chromatography and atomic emission detection. Journal of Chromatography A 1216(39):6735–6740. doi:10.1016/j.chroma.2009.08.019

Capriotti AL, Cavaliere C, Giansanti P, Gubbiotti R, Samperi R, Lagana A (2010) Recent developments in matrix solid-phase dispersion extraction. J Chromatogr A 1217(16):2521–2532. doi:10.1016/j.chroma.2010.01.030

Chen SJ, Cheng FC, Jen JF (2010) On-line microdialysis coupled solid-phase extraction to decrease matrix interference in the HPLC analysis of urinary ketamine and its metabolites. J Sep Sci 33(13):2010–2016. doi:10.1002/jssc.201000002

Cheng CY, Wang YC, Ding WH (2011) Determination of triclosan in aqueous samples using solid-phase extraction followed by on-line derivatization gas chromatography-mass spectrometry. Anal Sci 27(2):197–202

Covaci A, Mashiya JG, Molope I, Tshiame I, Molatlhegi R, Ngobeni P (2010) Miniaturized method based on matrix solid-phase dispersion for the rapid screening of 36 pesticides in agricultural food commodities. Anal Lett 43(7–8):1400–1410. doi:10.1080/00032710903518807

Cunha SS, Fernandes JO, Oliveira MBPP (2006) Quantification of free and esterified sterols in Portuguese olive oils by solid-phase extraction and gas chromatography-mass spectrometry. J Chromatogr A 1128(1–2):220–227. doi:10.1016/j.chroma.2006.06.039

De LGuardiaM, Garrigues S (2011) Challenges in green analytical chemistry. RSC Publishing, Cambridge

De Villiers A, Vanhoenacker G, Lynen F, Sandra P (2004) Stir bar sorptive extraction-liquid desorption applied to the analysis of hop-derived bitter acids in beer by micellar electrokinetic chromatography. Electrophoresis 25(4–5):664–669. doi:10.1002/elps.200305758

Duan C, Shen Z, Wu D, Guan Y (2011) Recent developments in solid-phase microextraction for on-site sampling and sample preparation. TrAC Trends Anal Chem 30(10):1568–1574

Eom IY, Tugulea AM, Pawliszyn J (2008) Development and application of needle trap devices. J Chromatogr A 1196:3–9. doi:10.1016/j.chroma.2008.02.090

Fan WL, Shen HY, Xu Y (2011) Quantification of volatile compounds in Chinese soy sauce aroma type liquor by stir bar sorptive extraction and gas chromatography-mass spectrometry. J Sci Food Agr 91(7):1187–1198. doi:10.1002/Jsfa.4294

Franc C, David F, Revel G de (2009) Multi-residue off-flavour profiling in wine using stir bar sorptive extraction-thermal desorption-gas chromatography-mass spectrometry. J Chromatogr A 1216(15):3318–3327. doi:10.1016/j.chroma.2009.01.103

Garcia-Lopez M, Canosa P, Rodriguez I (2008) Trends and recent applications of matrix solid-phase dispersion. Anal Bioanal Chem 391(3):963–974. doi:10.1007/s00216-008-1898-y

Gilbert-Lopez B, Garcia-Reyes JF, Lozano A, Fernandez-Alba AR, Molina-Diaz A (2010) Large-scale pesticide testing in olives by liquid chromatography-electrospray tandem mass spectrometry using two sample preparation methods based on matrix solid-phase dispersion and QuEChERS. J Chromatogr A 1217(39):6022–6035. doi:S0021-9673(10)00998-2 [pii], 10.1016/j.chroma.2010.07:062

Gonzalez-Marino I, Quintana JB, Rodriguez I, Schrader S, Moeder M (2011) Fully automated determination of parabens, triclosan and methyl triclosan in wastewater by microextraction by packed sorbents and gas chromatography-mass spectrometry. Anal Chim Acta 684(1–2):59–66. doi:10.1016/j.aca.2010.10.049

Gonzalez-Ruiz V, Olives AI, Martin MA (2011) SPE/RP-HPLC using C1 columns: an environmentally friendly alternative to conventional reverse-phase separations for quantitation of beta-carboline alkaloids in human serum samples. Anal Bioanal Chem 400(2):395–401. doi:10.1007/s00216-010-4562-2

Greenwood R, Mills GA, Roig B (2007) Introduction to emerging tools and their use in water monitoring. Trac-Trends Anal Chem 26(4):263–267. doi:10.1016/j.trac.2006.12.006

Guerrero ED, Bastante MJC, Mejias RC, Marin RN, Barroso CG (2011) Characterization and differentiation of Sherry Brandies using their aromatic profile. J Agr Food Chem 59(6):2410–2415. doi:10.1021/Jf104409n

Hampel D, Mosandl A, Wust M (2005) Biosynthesis of mono- and sesquiterpenes in carrot roots and leaves (Daucus carota L.): metabolic cross talk of cytosolic mevalonate and plastidial methylerythritol phosphate pathways. Phytochemistry 66(3):305–311. doi:10.1016/j.phytochem.2004.12.010

Hsu CL, Ding WH (2009) Determination of low-molecular-weight dicarboxylic acids in atmospheric aerosols by injection-port derivatization and gas chromatography-mass spectrometry. Talanta 80(2):1025–1028. doi:10.1016/j.talanta.2009.07.047

http://www.nemi.gov/

http://www.epa.gov/sw-846/pdfs/3510c.pdf

http://www.epa.gov/sw-846/pdfs/3540c.pdf

Hylton K, Mitra S (2007) Automated, on-line membrane extraction. J Chromatogr A 1152(1–2):199–214. doi:10.1016/j.chroma.2006.12.047

Hyotylainen T, Riekkola ML (2005) Solid-phase extraction or liquid chromatography coupled on-line with gas chromatography in the analysis of biological samples. J Chromatogr B Analyt Technol Biomed Life Sci 817(1):13–21. doi:S1570023204003964 [pii], 10.1016/j.jchromb.2004.04:033

Iparraguirre A, Prieto A, Navarro P, Olivares M, Fernandez LA, Zuloaga O (2011) Optimisation of stir bar sorptive extraction and in-tube derivatisation-thermal desorption-gas chromatography-mass spectrometry for the determination of several endocrine disruptor compounds in environmental water samples. Anal Bioanal Chem 401(1):339–352. doi:10.1007/s00216-011-5074-4

Insa S, Antico E, Ferreira V (2005) Highly selective solid-phase extraction and large volume injection for the robust gas chromatography-mass spectrometric analysis of TCA and TBA in wines. J Chromatogr A 1089(1–2):235–242. doi:10.1016/j.chroma.2005.06.061

Ishizaki A, Saito K, Hanioka N, Narimatsu S, Kataoka H (2010) Determination of polycyclic aromatic hydrocarbons in food samples by automated on-line in-tube solid-phase microextraction coupled with high-performance liquid chromatography-fluorescence detection. J Chromatogr A 1217(35):5555–5563. doi:S0021–9673(10)00868-X [pii], 10.1016/j.chroma.2010.06:068

Jakubowska N, Polkowska Z, Namiesnik J, Przyjazny A (2005) Analytical applications of membrane extraction for biomedical and environmental liquid sample preparation. Crit Rev Anal Chem 35(3):217–235. doi:10.1080/10408340500304032

Juan-Garcia A, Manes J, Font G, Pico Y (2004) Evaluation of solid-phase extraction and stir-bar sorptive extraction for the determination of fungicide residues at low-mu g kg(-1) levels in grapes by liquid chromatography-mass spectrometry. J Chromatogr A 1050(2):119–127. doi:10.1016/j.chroma.2004.08.026

Jochmann MA, Kmiecik MP, Schmidt TC (2006) Solid-phase dynamic extraction for the enrichment of polar volatile organic compounds from water. J Chromatogr A 1115(1–2):208–216. doi:10.1016/j.chroma.2006.02.061

Kanimozhi S, Basheer C, Narasimhan K, Liu L, Koh S, Xue F, Choolani M, Lee HK (2011) Application of porous membrane protected micro-solid-phase-extraction combined with gas chromatography-mass spectrometry for the determination of estrogens in ovarian cyst fluid samples. Anal Chim Acta 687(1):56–60. doi:10.1016/j.aca.2010.12.007

Kaljurand M, Koel M (2011) Recent advancements on greening analytical separation. Cri Rev Anal Chem 41(1):2–20. doi:Pii 933412012, 10.1080/10408347.2011.539420

Kawaguchi M, Ishii Y, Sakui N, Okanouchi N, Ito R, Inoue K, Saito K, Nakazawa H (2004) Stir bar sorptive extraction with in situ derivatization and thermal desorption-gas chromatography-mass spectrometry in the multi-shot mode for determination of estrogens in river water samples. J Chromatogr A 1049(1–2):1–8. doi:10.1016/j.chroma.2004.08.013

Keith LH, Gron LU, Young JL (2007) Green analytical methodologies. Chem Rev 107(6):2695–2708. doi:10.1021/Cr068359e

Klein DR, Flannelly DF, Schultz MM (2010) Quantitative determination of triclocarban in wastewater effluent by stir bar sorptive extraction and liquid desorption-liquid chromatography-tandem mass spectrometry. J Chromatogr A 1217(11):1742–1747. doi:10.1016/j.chroma.2010.01.028

Kloskowski A, Chrzanowski W, Pilarczyk M, Namiesnik J (2007) Modern techniques of sample preparation for determination of organic analytes by gas chromatography. Crit Rev Anal Chem 37(1):15–38. doi:10.1080/10408340600976663

Kouzayha A, Al Iskandarani M, Mokh S, Rabaa AR, Budzinski H, Jaber F (2011) Optimization of a solid-phase extraction method using centrifugation for the determination of 16 polycyclic aromatic hydrocarbons in water. J Agric Food Chem 59(14):7592–7600. doi:10.1021/Jf200123v

Koel M, Kaljurand M (2006) Application of the principles of green chemistry in analytical chemistry. Pure Appl Chem 78(11):1993–2002. doi:10.1351/pac200678111993

Kreck M, Scharrer A, Bilke S, Mosandl A (2002) Enantioselective analysis of monoterpene compounds in essential oils by stir bar sorptive extraction (SBSE)-enantio-MDGC-MS. Flavour Frag J 17(1):32–40

Kreck M, Puschel S, Wust M, Mosandl A (2003) Biogenetic studies in syringa vulgaris L.: synthesis and bioconversion of deuterium-labeled precursors into lilac aldehydes and lilac alcohols. J Agric Food Chem 51(2):463–469

Kristenson EM, Ramos L, Brinkman UAT (2006) Recent advances in matrix solid-phase dispersion. Trac-Trends Anal Chem 25(2):96–111

Kronholm J, Hartonen K, Riekkola ML (2007) Analytical extractions with water at elevated temperatures and pressures. Trac-Trends Anal Chem 26(5):396–412. doi:10.1016/j.trac.2007.03.004

Majors RE (1991) An overview of sample preparation. Lc Gc-Mag Sep Sci 9(1):16–18

Melo LP, Queiroz RH, Queiroz ME (2011) Automated determination of rifampicin in plasma samples by in-tube solid-phase microextraction coupled with liquid chromatography. J Chromatogr B Analyt Technol Biomed Life Sci 879(24):2454–2458. doi:S1570-0232(11)00438-7 [pii], 10.1016/j.jchromb.2011.06:041

Mieth M, Kischkel S, Schubert JK, Hein D, Miekisch W (2009) Multibed needle trap devices for on site sampling and preconcentration of volatile breath biomarkers. Anal Chem 81(14):5851–5857. doi:10.1021/Ac9009269

Moeder M, Bauer C, Popp P, Pinxteren M van, Reemtsma T (2012) Determination of pesticide residues in wine by membrane-assisted solvent extraction and high-performance liquid chromatography-tandem mass spectrometry. Anal Bioanal Chem 403(6):1731–1741. doi:10.1007/s00216-012-5956-0

Morlock G, Kopacz S (2008) Fast and precise SBSE-HPTLC/FLD method for quantification of six polycyclic aromatic hydrocarbons frequently found in water. J Liq Chromatogr R T 31(13):1925–1942. doi:10.1080/10826070802194856

Musshoff F, Lachenmeier DW, Kroener L, Madea B (2002) Automated headspace solid-phase dynamic extraction for the determination of amphetamines and synthetic designer drugs in hair samples. J Chromatogr A 958(1–2):231–238

Namiesnik J (2001) Green analytical chemistry—some remarks. J Sep Sci 24(2):151–153 do Nascimento I, Schossler P, Freitas LS, Melecchi MIS, Vale MGR, Caramao EB (2004) Selective extraction of benzoic acid from landfill leachate by solid-phase extraction and ion-exchange chromatography. J Chromatogr A 1027(1–2):167–170. doi:10.1016/j.chroma.2003.10.044

Navarro M, Pico Y, Marin R, Manes J (2002) Application of matrix solid-phase dispersion to the determination of a new generation of fungicides in fruits and vegetables. J Chromatogr A 968(1–2):201–209

Nevado JJB, Penalvo GC, Dorado RMR (2005) Method development and validation for the separation and determination of omeprazole enantiomers in pharmaceutical preparations by capillary electrophoresis. Anal Chim Acta 533(2):127–133. doi:10.1016/j.aca.2004.11.018

Ochiai N, Sasamoto K, Kanda H, Yamagami T, David F, Tienpont B, Sandra P (2005) Optimization of a multi-residue screening method for the determination of 85 pesticides in selected food matrices by stir bar sorptive extraction and thermal desorption GC-MS. J Sep Sci 28(9–10):1083–1092. doi:10.1002/jssc.200500017

Ochiai N, Sasamoto K, Kanda H, Nakamura S (2006) Fast screening of pesticide multiresidues in aqueous samples by dual stir bar sorptive extraction-thermal desorption-low thermal mass gas chromatography-mass spectrometry. Journal of Chromatography A 1130(1):83–90. doi:10.1016/j.chroma.2006.06.032

Ochiai N, Sasamoto K, Kanda H, Pfannkoch E (2008) Sequential stir bar sorptive extraction for uniform enrichment of trace amounts of organic pollutants in water samples. J Chromatogr A 1200(1):72–79. doi:10.1016/j.chroma.2008.05.069

Olbe L, Carlsson E, Lindberg P (2003) A proton-pump inhibitor expedition: the case histories of omeprazole and esomeprazole. Nat Rev Drug Discov 2(2):132–139. doi:10.1038/Nrd1010

Papadopoulos N, Gikas E, Zalidis G, Tsarbopoulos A (2009) Simultaneous Determination of herbicide terbuthylazine and its major hydroxy and dealkylated metabolites in typha latifolia L. Wetland plant using SPE and HPLC-DAD. J Liq Chromatogr R T 32(20):2975–2992. doi:Pii 916846758, 10.1080/10826070903320566

Pico Y, Fernandez M, Ruiz MJ, Font G (2007) Current trends in solid-phase-based extraction techniques for the determination of pesticides in food and environment. J Biochem Biophys Methods 70(2):117–131. doi:10.1016/j.jbbm.2006.10.010

Poole CF (2003) New trends in solid-phase extraction. Trac-Trend Anal Chem 22(6):362–373. doi:10.1016/S0165-9936(03)00605-8

Ponnusamy VK, Jen JF (2011) A novel graphene nanosheets coated stainless steel fiber for microwave assisted headspace solid phase microextraction of organochlorine pesticides in aqueous samples followed by gas chromatography with electron capture detection. J Chromatogr A 1218(39):6861–6868. doi:10.1016/j.chroma.2011.08.019

Popp P, Bauer C, Hauser B, Keil P, Wennrich L (2003) Extraction of polycyclic aromatic hydrocarbons and organochlorine compounds from water: a comparison between solid-phase microextraction and stir bar sorptive extraction. J Sep Sci 26(9–10):961–967. doi:10.1002/jssc.200301398

Prieto A, Basauri O, Rodil R, Usobiaga A, Fernandez LA, Etxebarria N, Zuloaga O (2010) Stir-bar sorptive extraction: a view on method optimisation, novel applications, limitations and potential solutions. J Chromatogr A 1217(16):2642–2666. doi:10.1016/j.chroma.2009.12.051

Qi DJ, Kang XJ, Chen LQ, Zhang YY, Wei HM, Gu ZZ (2008) Electrospun polymer nanofibers as a solid-phase extraction sorbent for the determination of trace pollutants in environmental water. Anal Bioanal Chem 390(3):929–938. doi:10.1007/s00216-007-1747-4

Ramos L (2012) Critical overview of selected contemporary sample preparation techniques. J Chromatogr A 1221:84–98. doi:10.1016/j.chroma.2011.11.011

Ramos L, Ramos JJ, Brinkman UA (2005) Miniaturization in sample treatment for environmental analysis. Anal Bioanal Chem 381(1):119–140. doi:10.1007/s00216-004-2906-5

Raynie DE (2004) Modern extraction techniques. Anal Chem 76(16):4659–4664. doi:10.1021/Ac040117w

Ridgway K, Lalljie SPD, Smith RM (2007) Sample preparation techniques for the determination of trace residues and contaminants in foods. J Chromatogr A 1153(1–2):36–53. doi:10.1016/j.chroma.2007.01.134

Ridgway K, Lalljie SPD, Smith RM (2010) An alternative method for analysis of food taints using stir bar sorptive extraction. Anal Chim Acta 677(1):29–36. doi:10.1016/j.aca.2010.01.049
Rodil R, Moeder M (2008) Stir bar sorptive extraction coupled to thermodesorption-gas chromatography-mass spectrometry for the determination of insect repelling substances in water samples. J Chromatogr A 1178(1–2):9–16. doi:S0021-9673(07)02003-1 [pii], 10.1016/j.chroma.2007.11:051
Rodrigues SA, Caldas SS, Primel EG (2010) A simple; efficient and environmentally friendly method for the extraction of pesticides from onion by matrix solid-phase dispersion with liquid chromatography-tandem mass spectrometric detection. Anal Chim Acta 678(1):82–89. doi:10.1016/j.aca.2010.08.026
do Rosario PMA, Nogueira JMF (2006) Combining stir bar sorptive extraction and MEKC for the determination of polynuclear aromatic hydrocarbons in environmental and biological matrices. Electrophoresis 27(23):4694–4702. doi:10.1002/elps.200600210
Rubert J, Soler C, Manes J (2011) Evaluation of matrix solid-phase dispersion (MSPD) extraction for multi-mycotoxin determination in different flours using LC-MS/MS. Talanta 85(1):206–215. doi:10.1016/j.talanta.2011.03.046
Sanchez-Rojas F, Bosch-Ojeda C, Cano-Pavon JM (2009) A review of stir bar sorptive extraction. Chromatographia 69:S79–S94. doi:10.1365/s10337-008-0687-2
Saito K, Yagi K, Ishizaki A, Kataoka H (2010) Determination of anabolic steroids in human urine by automated in-tube solid-phase microextraction coupled with liquid chromatography-mass spectrometry. J Pharm Biomed Anal 52(5):727–733. doi:S0731-7085(10)00113-5 [pii], 10.1016/j.jpba.2010.02:027
Sandra P, Tienpont B, Vercammen J, Tredoux A, Sandra T, David F (2001) Stir bar sorptive extraction applied to the determination of dicarboximide fungicides in wine. J Chromatogr A 928(1):117–126
Sandra P, Tienpont B, David F (2003) Multi-residue screening of pesticides in vegetables, fruits and baby food by stir bar sorptive extraction-thermal desorption-capillary gas chromatography-mass spectrometry. J Chromatogr A 1000(1–2):299–309. doi:10.1016/S0021-9673(03)00508-9
Sasamoto K, Ochiai N, Kanda H (2007) Dual low thermal mass gas chromatography-mass spectrometry for fast dual-column separation of pesticides in complex sample. Talanta 72(5):1637–1643. doi:10.1016/j.talanta.2007.03.020
Serodio P, Nogueira JMF (2006) Considerations on ultra-trace analysis of phthalates in drinking water. Water Res 40(13):2572–2582. doi:10.1016/j.watres.2006.05.002
Sewenig S, Bullinger D, Hener U, Mosandl A (2005) Comprehensive authentication of (E)-alpha(beta)-ionone from raspberries, using constant flow MDGC-C/P-IRMS and Enantio-MDGC-MS. J Agric Food Chem 53(4):838–844. doi:10.1021/Jf040356k
Tan BLL, Hawker DW, Muller JF, Tremblay LA, Chapman HF (2008) Stir bar sorptive extraction and trace analysis of selected endocrine disruptors in water, biosolids and sludge samples by thermal desorption with gas chromatography-mass spectrometry. Water Res 42(1–2):404–412. doi:10.1016/j.watres.2007.07.032
Tholl D, Boland W, Hansel A, Loreto F, Rose US, Schnitzler JP (2006) Practical approaches to plant volatile analysis. Plant J 45(4):540–560. doi:TPJ2612 [pii], 10.1111/j.1365-313X.2005.02612.x
Tobiszewski M, Mechlinska A, Zygmunt B, Namiesnik J (2009) Green analytical chemistry in sample preparation for determination of trace organic pollutants. Trac-Trend Anal Chem 28(8):943–951. doi:10.1016/j.trac.2009.06.001
Tobiszewski M, Mechlinska A, Namiesnik J (2010) Green analytical chemistry-theory and practice. Chem Soc Rev 39(8):2869–2878. doi:10.1039/B926439f
Trevisan MGP, Poppi RJ (2006) Química analítica de processos. Química Nova 29(5):1065–1071
Turner C (2006) Overview of modern extraction techniques for food and agricultural samples. Acs Sym Ser 926:3–19
Vinas P, Campillo N, Martinez-Castillo N, Hernandez-Cordoba M (2009) Solid-phase microextraction on-fiber derivatization for the analysis of some polyphenols in wine and grapes using gas chromatography-mass spectrometry. J Chromatogr A 1216(9):1279–1284. doi:10.1016/j.chroma.2008.12.058

Vinas P, Campillo N, Martinez-Castillo N, Hernandez-Cordoba M (2010) Comparison of two derivatization-based methods for solid-phase microextraction-gas chromatography-mass spectrometric determination of bisphenol A, bisphenol S and biphenol migrated from food cans. Anal Bioanal Chem 397(1):115–125. doi:10.1007/s00216-010-3464-7

Wille SM, Lambert WE (2007) Recent developments in extraction procedures relevant to analytical toxicology. Anal Bioanal Chem 388(7):1381–1391. doi:10.1007/s00216-007-1294-z

Wilson ID, Brinkman UAT (2007) Hype and hypernation: multiple hyphenation of column liquid chromatography and spectroscopy. Trac-Trends Anal Chem 26(9):847–854. doi:10.1016/j.trac.2007.07.007

Wu F, Lu WP, Chen JH, Liu W, Zhang L (2010) Single-walled carbon nanotubes coated fibers for solid-phase microextraction and gas chromatography-mass spectrometric determination of pesticides in Tea samples. Talanta 82(3):1038–1043. doi:10.1016/j.talanta.2010.06.016

Yang TJ, Tsai FJ, Chen CY, Yang TCC, Lee MR (2010) Determination of additives in cosmetics by supercritical fluid extraction on-line headspace solid-phase microextraction combined with gas chromatography-mass spectrometry. Anal Chim Acta 668(2):188–194. doi:10.1016/j.aca.2010.04.001

Yamaguchi C, Lee WY (2010) A cost effective, sensitive, and environmentally friendly sample preparation method for determination of polycyclic aromatic hydrocarbons in solid samples. J Chromatogr A 1217(44):6816–6823. doi:10.1016/j.chroma.2010.08.055

Yu YQ, Gilar M, Kaska J, Gebler JC (2005) A rapid sample preparation method for mass spectrometric characterization of N-linked glycans. Rapid Commun Mass Spectrom 19(16):2331–2336. doi:10.1002/Rcm.2067

Zuin VG, Lopes AL, Yariwake JH, Augusto F (2004) Application of a novel sol-gel polydimethylsiloxane-poly(vinyl alcohol) solid-phase microextraction fiber for gas chromatographic determination of pesticide residues in herbal infusions. J Chromatogr A 1056(1–2):21–26. doi:10.1016/j.chroma.2004.07.074

Zuin VG, Montero L, Bauer C, Popp P (2005) Stir bar sorptive extraction and high-performance liquid chromatography-fluorescence detection for the determination of polycyclic aromatic hydrocarbons in Mate teas. J Chromatogr A 1091(1–2):2–10

Zuin VG, Schellin M, Montero L, Yariwake JH, Augusto F, Popp P (2006) Comparison of stir bar sorptive extraction and membrane-assisted solvent extraction as enrichment techniques for the determination of pesticide and benzo[a]pyrene residues in Brazilian sugarcane juice. J Chromatogr A 1114(2):180–187. doi:10.1016/j.chroma.2006.03.035

Zygmunt B, Jastrzebska A, Namiesnik J (2001) Solid phase microextraction—a convenient tool for the determination of organic pollutants in environmental matrices. Crit Rev Anal Chem 31(1):1–18

Chapter 8
Studies Regarding the Optimization of the Solvent Consumption in the Determination of Organochlorine Pesticides From Complex Fodder

Adriana Chiş, Purcărea Cornelia and Cristina Horga

Abstract Since the main pathway for animal food contamination by OCP is the ingestion of polluted feed, the determination of the contamination degree of the fodders is extremely important. In case of complex fodders in whose composition there are various materials, the potential contamination sources are multiple. The determination of organochlorine pesticides was made through the gas chromatography method using capillary columns and detector with electrons capture. For the separation step, a method specific to non-fat aliments was performed. This method requires the use of three subsequent elutions with 200 ml ethylic ether/light petroleum, in variable proportions. For the purpose of optimizing the method, there were determinations performed on the same samples through extraction in the normal variant and in reduced ones. This way, we could verify the level up to which the solvent consumption can be reduced without affecting the accuracy of the determination. The practical determinations proved that only the 1:2 reduced method has fully qualitative application and specific quantitative applications.

8.1 Introduction

The organochlorine pesticides (OCP in the text) take part of Persistent Organic Polutants (POPs) list from the beginning of their investigation by United Nations Environment Programme because of their resistance to degradation which is responsible of their persistence in the environment. In the European Union, the use of this type of pesticide was gradually decreased up to the total interdiction of the last product admitted (Endosulfane), starting with 2008. The residues are found at smaller and smaller concentration levels but it continues to be supervised at Europe-

A. Chiş (✉) · P. Cornelia
Department of Food control, Environment Protection Faculty,
University of Oradea, Oradea, Romania
e-mail: andichis@yahoo.com

C. Horga
Institute of Public Health, Cluj-Napoca, Romania

Inamuddin, A. Mohammad (eds.), *Green Chromatographic Techniques*,
DOI 10.1007/978-94-007-7735-4_8,

an level for reasons of food safety. Therefore, they must be detected (EFSA 2005a, b, 2006a, b) in food and feed in all European countries including Romania. These studies are conducted even in neighboring countries of Romania which are not subject to the European Union norms, such as Serbia. (Škrbič and Predojevič 2008).

Since the main pathway for animal food contamination by OCPs is the ingestion of polluted feed, the determination of the contamination degree of the fodders is extremely important. In case of complex fodders in whose composition there are various materials, the potential contamination sources are multiple.

While in European Union the use of OCPs for crops protection or silo treatments has been banned for more than 20 years, in Asia, South America or South Africa, some of them like DDT, are still in use especially for malaria control. This is why in this area there were many studies regarding the impact of the pesticide contamination on fodder or food on the humans' health state (Gill et al. 2001, 2009; Singh et al. 1997; Kannan et al. 1992). But the effect of using such compounds can spread on other areas as well, for various reasons. First of all, this way the pollution of the environment factors is maintained, a phenomenon which does not have boundaries (Wang et al. 2007). Moreover, there is the possibility that the complex fodders become impurified through a component coming from such an area, for example cereals where the presence of OCPs was highlighted (Bai et al. 2006).

The determination of non volatile chemical contaminants—OCP type—is performed in two phases: the preparation of the samples and the determination *per se*. The preparation phase consists of the separation of the toxic from the matrix while the purification of the extract is essential in what regards the results obtained. The same time, it represents the phase with the highest consumption of work time and expensive reagents. For this reason, the studies regarding the possibilities to optimize this method, without affecting the accuracy of the results obtained presents a real interest.

Since we know that the decrease of the concentration makes the analytic determination more difficult, we chose the application of a multiple-elution method, which has high chances to highlight all the compounds present, even at ppb concentrations (SR EN 12393-2/2003). However, at the same time, we wished to highlight the use of the three elutions practices according to the purpose followed.

8.2 Materials and Methods

8.2.1 Materials

For the study that makes the object of the present paper, two series of two different complex fodder samples have been subject to testing. The fodder, made of unique fodder mixes (Total Mixed Ratios) come from a cow factory in the area of the Sânandrei village, nearby the city of Oradea. The samples were coded TRM1 (1), (2) and TRM2 (1), (2).

TRM1 are combined fodders for high productivity lactating cows—that is between 25–40 l/day. They consist of a mixture of several components: silo alfalfa, beer marc, grounded corn, grounded triticale, PMV premixed (protein-vitamin-mineral concentrate and soy grist), calcium carbonate, sodium chloride. TRM2 are combined fodders for medium productivity lactating cows—that is between 10–25 l/day. They consist of a mixture made by the same components as TRM1, the difference consisting in the used quantity of cereals and PVM. We choose this types of complex fodders because we considered that through the bioaccumulation characteristic specific to OCPs, the possible residues in fodders might influence the value of these contaminants in milk and fat dairy products. This action is connected to the fact that these two types of fodders differ exactly through the cereal content and PVMs (Halga et al. 2005).

The samples were taken in plastic water-tight containers. The recipients used for the determination of OCP were rinsed with petroleum ether; after cleaning, they were dried in the drying oven and kept with the lid on in order to prevent contamination.

8.2.2 Methods

The Preparation of the Samples for the Analysis The preparation for the analysis consisted of the determination of their humidity because the OCP extraction technique from non-fat vegetal products depends on the water content of the samples (SR EN 12393-2/2003). The used method was the drying in thermo-adjustable oven at 105 °C, for 4 h (Bradley 2003); the determination was performed immediately after the collection, in the same day, in the Food Toxicology Laboratory of the Faculty of Environment Protection from Oradea.

The Preparation of the Samples for Chromatography The way this phase is performed influences seriously upon the results obtained, as in the case of fat products. The extraction technique recommended for wet products is to mix the minced sample with the solvent, in order to obtain a maximum efficiency of POC residue extraction and a minimum amount of co-extracted substances which can show the risk of interference in determination (Anastassiades et al. 2005). The solvent extraction is followed by the liquid/liquid partition. The extract is purified on column with activated Florisil®. The OCPs are eluted with ethylic ether/light petroleum in variable proportions. The extraction and the determination of the contaminants were performed on fresh products in the following 48 hours after the collection, at the Institute of Public Health from Cluj Napoca. All the reagents used were Merck-type with chromatographic purity and there were no additional purifications performed, except for witness samples of the reagents on each work phase.

The first phase undertaken is the extraction of the toxic with an extraction mixture of acetonitryl and water (65:35 V/V), because the humidity was under 75 %. The resulting filtrate (F) is extracted through partition with light petroleum, in the separation funnel. The extract (P) is passed to the purification phase; the F and P values in the phase of sample preparation are written during the determination because they are used in the calculation of the sample amount that passes through the purification column. There follows the purification of Florisil column with three subsequent elutions of 200 ml mixture ethylic ether/ light petroleum for each, in the following V/V proportions: 6:94 (eluent A), 15:85 (eluent B) and 50:50 (eluent C). Further on this paper, letters A, B and C refers at the specific eluent above nominated. This kind of determination require the use of organic solvent because the OCPs are founded in the grease of the food witch need to be solved in order to perform the determination. So the "green" approach consist in founding ways to diminish the consumption of this chemicals. As one can see, the solvent consumption, for a single sample is consistent. For this reason, the method has been tested on the same samples both in the 1:1 variant and in reduced variants: 1:2, respectively 1:5. The reduction referred proportionally to the amounts of sample used, the amounts of solvents used in all the work phases and the dimension of the glassware used.

The Determination of the OCP Residues in the Studies Products We have used the method most recommended by the specialised literature (Tadeo 2008; Hura 2006), as set forth by current legislation for the calculation of pesticide residues in vegetal products. It is the gas chromatography (SR EN 12393-3/2003) using capillary column. The GC technique on capillary column has a higher separation power than the one with filling column and that is why they are especially recommended for complex extracts, such as the case of determinations of multiresidues of organochlorinated pesticides. (Koerner 2001).

Used Devices We have used a GC 2010 Shimadzu gas chromatograph, with the following characteristics:

- Capillary column type RTX -CL- pesticides 30 m length an 0.25 mm diameter. The column works at a temperature between 150÷320 °C with a gradient of 3 °C/2 min
- Detector with electrons capture (ECD), nuclid ^{63}Ni—370 MBq (10 mCu)
- Autosampler injection system with 6+2 spaces for vials, type AOC-210

Chromatography conditions:

- Injection temperature (splitting) = 250 °C
- Splitting pressure = 163.5 Kpa
- Splitting gas: He with a 124 ml/min flow at scavenging 30 ml/min
- Carrying gas: N_2 ultrapure 99.99 %
- Detector: T = 320 °C, detector current of 2 nA, make-up flow = 30 ml/min

Qualitative and Quantitative Calculations For the qualitative and quantitative calculation of the contaminants possibly present in the tested products, we used a standard produces by the RESTEK company No 32292, Lot nr A021837, type "Organochlorine pesticide Mix AB≠2" having a concentration of 200 ppb. Notice the fact that the standard contains, with one exception (hexachlorobenzene) all the organochlorine compounds under sanitary-veterinary surveillance in vegetal and animal foods. The standard was used at the 50 ppb dilution.

Complex fodder tests submitted to the verification from the point of view of the contamination with OCPs have undergone the procedures explained in "Test preparation for chromatography" part of the paper. The extract purified, retaken in petroleum ether was submitted to the chromatography under the same conditions as the standard test, as well as the blanks-test of the used reagents, according to the separation/purification method.

For the quantitative determinations, we have used the value of the surface of the drops (compounds that have been previously identified as being present in the tested sample). In order to prevent possible calculation errors, we introduced the data we obtained through the automatic integration of the unknown tests and the used standard in EXCELL calculation sheets, selecting only the surface of the picks we need, that is the compounds that were found from the qualitative point of view.

In the case of the non-fat products (fodders) the calculation of the OCPs residues concentration refers to the amount of sample that passes through the Florisil purification column (S). The S value is calculated according to the moisture of the products and using parameters that result from the development of all the work procedure for non-fat products, as explain in equation (8.1) (SR EN 12393-2/2003).

$$S = (m \times F \times P)/T \times 100, \ g \tag{8.1}$$

where:

S: the amount of sample, g, passed through the purification column
m: the mass of the sample, g
F: the measured volume of the filtrate after the acetonytril extraction, ml
T: total volume, ml; (ml water from the sample plus ml of acetonytril added, less the empiric contraction volume;
P the measured extract of light petroleum, ml
100 the volume of the light petroleum used in the extraction, ml

In the case of non-fat products, the values obtained are not corrected with the recovery degree (SR EN 12393, 3-2004).

8.3 Results

For the qualitative determination, we compared the retention times for the significant picks that appear on the chromatogram of the unknown tests with the ones of the compounds from the standard sample. This way, we can determine which of the compounds that are present in the standard can also be found in the tested sample (Gocan 1998).

Table 8.1 Caracterisation of the mixt OCP standard

A	OCP in the standard	B
1	α HCH	7.194
2	γ HCH	8.426
3	β HCH	8.886
4	δ HCH	9.552
5	Heptaclor	10.372
6	Aldrin	11.657
7	Heptaclor epoxid	14.652
8	γ Chlordan	15.274
9	α Chlordan	15.951
10	4,4' DDE	16.518
11	α endosulfan	16.753
12	Dieldrin	17.766
13	Endrin	18.878
14	4,4' DDD	19.853
15	β Endosulfan	20.076
16	4,4' DDT	21.305
17	Endrin aldehida	22.385
18	Metoxiclor	24.346
19	Sulfat endosulfan	24.693
20	Endrin cetona	26.030

A Elution order, *B* Retention time in the standard

The results are written as follows:

- Table 8.1 shows the elution order and retention times for each component of the used standard. The tables containing the results, both qualitative and quantitative, will have only the number afferent to the detected OCP.
- TRM1: Table 8.2, 8.3 and 8.4, for the three variants applied separately for the three elution solvents used.
- TRM2: Table 8.5, 8.6 and 8.7, for the three variants applied separately for the three elution solvents used.

For the quantitative determinations, it was calculated the amounts of samples passed through the purification column, in the three work variants applied to the tested samples. The values are found in Table 8.8.

The results of the quantitative determination are shown in Tables 8.9 and 8.10, for the three work variants applied.

The concentration written in the table represent the amount of the values at the three elutions applied, if case. The expression of the concentration as isomer sum for α and β HCH, endosulfan and DDT complies with the European laws (EFSA 2005a, 2005b, 2006a, 2006b) and the national laws (Legis).

Table 8.2 Qualitative calculation of OCP residues, TMR1 complex fodder, variant 1:1

A	Retention time in the sample, eluent A, B and C					
	TRM1(1)			TRM1(2)		
	E_A	E_B	E_C	E_A	E_B	E_C
1	7.181	–	–	7.179	–	–
2	8.418	–	–	8.411	8.422	–
3	8.873	–	8.845	–	–	8.864
5	10.375	–	–	10.366	–	–
6	11.656	–	–	11.662	–	–
7	14.64	–	–	14.640	–	–
8	15.261	15.289	–	–	–	–
9	15.942	–	–	16.941	–	–
10	16.538	–	–	16.525	–	–
11	17.748	–	16.766	17.745	–	–
12	17.766	17.763	–	–	–	–
13	18.866	18.867	–	18.856	18.877	–
14	19.848	–	–	19.831	–	–
15	–	20.068	–	–	–	–
16	21.286	–	–	21.282	–	–
19	24.816	–	–	24.801	–	–

Table 8.3 Qualitative calculation of the OCP residues, TRM1 complex fodder, variant 1:2

A	Retention time in the sample, eluent A, B and C					
	TRM1(1)			TRM1(2)		
	E_A	E_B	E_C	E_A	E_B	E_C
1	7.188	–	–	7.184		
2	8.419	–	–	8.433	8.368	
3	8.857	–	–		8.861	
5	10.368	–	–	10.359	–	–
6	11.65	–	–	11.65	–	–
7	14.637	–	–		–	–
8	15.261	15.286	–	–	–	–
9	15.938	–	–	15.945	–	–
10	16.563	–	–	16.528	–	–
11	17.741	–	–	17.751	–	–
12	17.753	–	–		–	–
13	18.849	–	–	18.866	18.976	–
14	19.834	–	–	19.839	–	–
15	–	20.06	–	–	–	–
16	21.285	–	–	21.278	–	–
19	24.812	–	24.675	24.767	–	–

Table 8.4 Qualitative calculation of the OCP residues, TRM1 complex fodder, variant 1:5

A	Retention time in the sample, eluent A, B and C					
	(1)			(2)		
	E_A	E_B	E_C	E_A	E_B	E_C
1	7.179	–	–	7.191	–	–
2	8.407	8.408	–	8.418	8.411	–
3	8.946	–	–	8.955	–	–
10	16.577	–	–	16.556	–	–
11	16.733	–	16.721	16.742	–	16.721
13	18.837	18.839	–	18.873	18.839	–
15	–	20.039	–	–	–	–
16	21.279	21.272	–	–	21.278	–
19	–	–	24.658	–	–	24.678

Table 8.5 Qualitative calculation of OCP residues, TMR2 complex fodder, variant 1:1

A	Retention time in the sample, eluent A, B and C					
	TRM2(1)			TRM2(2)		
	E_A	E_B	E_C	E_A	E_B	E_C
1	7.189	–	–	7.171	–	–
2	8.420	–	–	8.416	–	–
3	8.859	–	8.841	8.844	–	8.852
5	10.362	–	–	10.367	–	–
6	11.649	–	–	11.652	–	–
7	13.653	–	–	14.650	–	–
8	15.272	15.268	–	–	–	–
9	15.933	–	–	15.946	–	–
10	16.534	–	–	16.531	–	–
11	16.744		16.780	16.740	–	16.768
12	17.758	–	–	17.761	–	–
13	18.857	18.967	–	18.866	18.872	
14	19.834	–	–	19.824	–	–
16	21.288	–	–	21.290	–	–

Table 8.6 Qualitative calculation of the OCP residues, TRM2 complex fodder, variant 1:2

A	Retention time in the sample, eluent A, B and C					
	(1)			(2)		
	E_A	E_B	E_C	E_A	E_B	E_C
1	7.182	–	–	7.177	–	–
2	8.412	–	–	8.408	8.431	–
3	8.851	–	–	8.864	–	–
5	10.361	–	–	10.358	–	–
6	11.645	–	–	11.671	–	–
7	13.663	–	–	13.663	–	–
8	15.268	15.265	–	–	–	–
9	15.934	–	–	15.948	–	–
10	16.567	–	–	16.544	–	–
11	16.739	–	16.842	16.744	–	16.823
12	17.751	–	–	17.759	–	–
13	18.848	18.979	–	18.855	18.968	–
14	19.834	–	–	19.851	–	–
16	21.285	–	–	21.292	–	–

Table 8.7 Qualitative calculation of the OCPs residues, TRM2 complex fodder, variant 1:5

A	Retention time in the sample, eluent A, B and C					
	(1)			(2)		
	E_A	E_B	E_C	E_A	E_B	E_C
1	–	–	–	7.182	–	–
2	8.419	8.408	–	8.421	8.413	–
3	8.791	–	–	8.852	–	–
11	16.737	–	–	16.744	–	–
13	18.855	18.867	–	18.865	18.890	–
16	21.288	–	21.404	21.311	–	–
19	24.622	–	–	–	–	–

A Elution order; E_A, E_B, E_C Elution solvents used: A, B, C

Table 8.8 The sample amount submitted to purification, complex fodder, variant 1:1, 1:2 and 1:5

Sample, variant		U, %	M, g	F, ml	P, ml	Water, ml	T, ml	S, g
TRM1	1:1	64.3	25.34	330	87	16.29	361.29	20.14
(1)	1:2		13.82	158	42	8.89	181.39	10.11
	1:5		6.04	62	15.5	3.88	72.88	3.98
TRM1	1:1	63.3	29.7	328	86	18.80	363.80	23.03
(2)	1:2		11.76	157	41	7.44	179.94	8.41
	1:5		5.77	63	17	3.65	72.65	4.25
TRM2	1:1	69.9	27.48	340	89	19.21	364.21	22.83
(1)	1:2		12.96	159	42	9.06	181.56	9.53
	1:5		5.3	56	16.5	3.70	72.70	3.37
TRM2	1:1	67.8	27.52	337	87	18.66	363.66	22.19
(2)	1:2		10.25	155	40	6.95	179.45	7.08
	1:5		5.45	65	17.5	3.70	72.70	4.26

Table 8.9 The quantitative calculation of the OCP residues, TRM 1 complex fodder

Nr*	TRM1(1)			TRM1(2)		
	Work variant					
	1:1	1:2	1:5	1:1	1:2	1:5
1	0.0049	0.0040	0.0017	0.0022	0.0018	0.0010
3						
2	0.0107	0.0077	0.0029	0.0131	0.0098	0.0051
5	0.0018	0.0033	–	0.0011	–	–
7						
6	0.0040	0.0040	–	0.0021	0.0016	–
8	0.0062	0.0057	–	0.0031	0.0021	–
9						
11						
15	0.0133	0.0096	0.0060	0.0114	0.0102	0.0048
19						
13	0.0028	0.0005	0.0011	0.0013	0.0007	0.0011
10						
14	0.0105	0.0049	0.0038	0.0128	0.0091	0.0044
16						

Table 8.10 The quantitative calculation of the OCP residues, TRM 2 complex fodder

Nr*.	TRM2(1)			TRM2(2)		
	Work variant					
	1:1	1:2	1:5	1:1	1:2	1:5
1	0,0022	0.0024	0.0003	0,0012	0.0013	–
3						
2	0,0038	0.0027	0.0021	0,0031	0.0019	0.0019
5	0,0011	0.0015	–	0,0008	0.0005	0.0022
7						
6	0,001	0.0013	–	0,0012	0.0014	–
8	0,0017	0.0021	–	0,0006	0.0005	–
9						
11						
15	0,0079	0.0074	0.0052	0,0048	0.0041	–
19						
13	0,0041	0.0031	0.0022	0,0063	0.0055	0.0026
10						
14	0,0069	0.0056	0.0025	0,0056	0.0035	0.0041
16						

8.4 Discussions

The discussions consist of the comparison of the results obtained in the normal 1:1 work variant, with the ones obtained in the two reduces variants applied—1:2, respectively 1:5 for the two types of complex fodder being tested. The comparison refers both at the types of OCPs revealed and at the value of the concentration of the residues identified quantitavely.

8.4.1 TRM1

The Comparison of the 1:1 Variants with the 1:2 Variant From the qualitative point of view, it is noteworthy that the same types of organochlorinated types of compounds have been identified through both methods, respectively the λ, β and γHCH isomers, Heptaclor isomers, Aldrin, Chlordan isomers, Dieldrin, Endosulfan isomers, Endrin, DDD isomers all the tested samples (Tables 8.2 and 8.3) in eluent A which we already reported seem to be enough for routine determinations (Chiş et al. 2011). The only exception is Heptaclor epoxid who doesn't appear in eluent A for (2) series. There are differences for the C eluent but they refers to compound appearing in others eluent, so the qualitative results are not affected.

From the quantitative point of view (Table 8.9) the values obtained through the 1:2 method depends on the evaluated OCP compound. So, this values, comparated with those from 1:1 variant, range between:

- α and β HCH isomers: 81–82 %

- γ HCH: 72–75 %
- Heptachlor isomers: not appear—183 %
- Aldrin: 76–100 %
- Chlordan isomers: 68–92 %
- Endosulphane isomers: 72–89 %
- Endrin: 18–54 %
- DDT isomers: 47–71 %

As it can be seen, for α and βHCH isomers the results are close and homogeny in the two series of tested samples. For γHCH the results are homogeny but the difference is important because the absolute values obtained are close to the European MRL for this compound that is 0.01 ppm.

For Aldrin, Chlordan isomers and Endosulphane isomers the obtained values are close but not homogeny in series (1) and (2). As for Heptachlor isomers, Endrin and DDT isomers the values are also not homogene but far-off also.

The Comparison of the 1:1 Variants with the 1:5 Variant From the qualitative point of view, it is noteworthy the fact that between the two variants there are relevant differences regarding the found compounds and the way they are eluted (Tables 8.2 and 8.4).

So, Heptachlor isomers, Aldrin, Chlordan isomers, Dieldrin and 4,4'DDD are not to be found both in series (1) and (2). Moreover, the one founded appear differently: endosulphane sulphate in eluent C instead of eluent A (series 1 and 2) and βHCH in eluent A instead of eluent C (series 2).

From the quantitative point of view (Table 8.9) the differences are noteworthy for the compounds that can be found quantitavely and for which the integrated area is enough to allow the calculation. So this values (series 1 and 2), comparated with those from 1:1 variant, range between:

- α and β HCH isomers: 35–45 %
- γ HCH: 27–39 %
- Endosulphane: 42–45 %
- Endrin: 39–85 %
- DDT isomers: 34–36 %

With one exception, Endrin in series (1), all the values are under 45 % from those calculated in the 1:1 variant.

8.4.2 TRM2

The Comparison of the 1:1 Variants with the 1:2 Variant From the qualitative point of view, the results are very much similar as those for TRM1. So, the same types of organochlorinated types of compounds have been identified through both methods, respectively the λ, β and γHCH isomers, Heptaclor isomers, Aldrin, Chlordan isomers, αEndosulfan, Dieldrin, Endrin, DDD isomers all the tested samples.

(Tables 8.5 and 8.6) in eluent A. The differences between the two variants appear for B and C elutions but the qualitative results are not affected because it refers to compound who was already detected in A eluent.

From the quantitative point of view (Table 8.10) the values obtained through the 1:2 method depends on the evaluated OCP compound. So this values, compared with those from 1:1 variant, range between (series 1 and 2):

- α and β HCH isomers: 108–109 %
- γ HCH: 68–71 %
- Heptachlor isomers: 62–136 %
- Aldrin and Dieldrin: 116–130 %
- Chlordan isomers: 83–123 %
- Endosulphane isomers: 85–94 %
- Endrin: 76–87 %
- DDT isomers: 62–81 %

In the case of TRM2 there are more OCP compounds presenting close and homogeny values: α, β and γHCH isomers, Aldrin and Dieldrin, Endosulphane isomers and Endrin.

For Chlordan isomers the obtained values are close but not homogeny in series (1) and (2). As for Heptachlor isomers and DDT isomers the values are also not homogene but far-off also in series 2.

All the calculated values are far away from the respective MRL, at least one order of magnitude.

The Comparison of the 1:1 Variants with the 1:5 Variant The situation is very much similar that with TRM1, namely between the two variants there are relevant differences from the point of view of the found compound and the way they are eluted (Tables 8.5 and 8.7).

So, Heptachlor isomers, Aldrin, Chlordan isomers, Dieldrin and 4,4'DDD are not to be found both in series (1) and (2) and αHCH in series 1. Moreover, eluent C has only a signal for 4,4'DDT in series 1 and eluent B a signal for γHCH witch doesn't exist in 1:1 variant.

From the quantitative point of view (Table 8.10) the differences are noteworthy for the compounds that can be found quantitavely and for which the integrated area is enough to allow the calculation. So this values (series 1 and 2), comparated with those from 1:1 variant, range between:

- α and β HCH isomers: not appear—14 %
- γ HCH: 55–61 %
- Endosulphane: not appear—65 %
- Endrin: 41–53 %
- DDT isomers: 36–73 %

With the exception of γ, Endrin in series (1), all the values are under 45 % from those calculated in the 1:1 variant. The values are not homogenous and with the exception of DDT, all the values are under 65 % from those calculated in the 1:1 variant.

8.5 Conclusions

In the determination of the organichlorinated pesticide residues the 1:2 reduced variant has very good qualitative application for both types of tested complex fodder, like for green fodders previousely tested (Chiş 2010).

As for the quantitative applications, those seem to be confined depending on purpose of the determination and the composition of the tested fodder.

For monitoring purpose of all OCP compounds this variant can be used if the values are not close to the admitted MRL in with case the 1:1 variant should be applied for confirmation.

For research purpose, concerning specific types of OCP there are differences between the OCP compounds. Thus for HCH isomers, Aldrin and Dieldrin, Chlordan isomers Endosulphane isomers the 1:2 variant can be used for both tested fodders. For Endrin, the different composition of the fodder affects the results. As for heptachlor isomers and DDT isomers, the results are inconclusive. Those conclusions are different from those obtained for green fodder where the reduced variant 1:2 has fully quantitative applications (Chiş, 2010), so it is obvious that the experiments must be done on specific types of fodder. The more the composition is complex, the more reduced variant must be applied with caution.

On the other hand, the 1:5 variant does not have a qualitative or quantitative applicability neither for TRM1 or TRM2. This thing can be put on behalf of the solvent and sample amount used, which prove to be insufficient in the field of 10^{-3}–10^{-4} ppm concentrations.

In conclusion only the 1:2 reduced method has fully qualitative application and specific quantitative applications but even so the economy of solvent is significant, moreover when a large number of samples has to be tested. In the same time, the duration of the determinations is shortened without affecting the accuracy, by maintaining the maximum elution speed at 5 ml/min.

References

Anastassiades M, Mastovska K, Lehotay S (2005) Study of a combination of analytic protectants to overcome matrix effects in routine GC analysis of pesticides residues in food matrices. Anal Chem 77(24):129–137

Bai Y, Zhou L, Li J (2006) Organochlorine pesticide (HCH and DDT) residues in dietary products from Shaanxi province, People's Republic of China. Bull Environ Contam Toxicol 76:422–446

Bradley RL Jr (2003) Moisture and total solid analysis, Chap. 6, in Part II, Food Analysis, 3rd edn. Springer Science (Suzanne Nielsen Ed.), USA Kluwer Academic New York, pp 81–103

Chiş A (2010) Studies regarding the optimization of the solvent consumption in the determination of organochlorine pesticides from green foder. Studia Universitatis "Vasile Goldis". Life Sci Ser 20:81–87

Chiş A, Purcărea C, Horga C (2011) The purpose of the multiple elutions for the determination of organochlorine pesticides in complex fodder. Studia Universitatis "Vasile Goldis". Life Sci Ser 21(1):87–94

EFSA (2005a) Opinion of the scientfic panel on contaminants in the food chain on a request from the commission related to gamma—HCH and other hexachlorocyclohexanes as undesirable substances in animal feed. EFSA J 250:1–39

EFSA (2005b) Opinion of the scientfic panel on contaminants in the food chain on a request from the commission related to Endrin as undesirable substances in animal feed. EFSA J 286:1–31

EFSA (2006a) Opinion of the scientfic panel on contaminants in the food chain on a request from the commission related to Endosulfan as undesirable substances in animal feed. EFSA J 234:1–31

EFSA (2006b) Opinion of the scientfic panel on contaminants in the food chain on a reques tfrom the commission related to DDT as undesirable substances in animal feed. EFSA J 433:1–69

Gill J, Aulakh RS, Mundi JPS, Sharma JK (2001) Pesticides residues in animal feed: a potential source of public health hazard. Paper in Post Conference proceeding of International Conference on Pesticides, Environment, Food Security from 19–23, November 2001 held at New Delhi. pp 160–163

Gill J, Sharma JK, Aulakh RS (2009) Studies on organochlorine pesticide residues in butter in Punjab. Toxicol Int 16:133–136

Gocan S (1998) High-performance cromatography, part I, gas chromatography. Dacia Publishing House, Cluj Napoca

Halga P, Pop IM, Avarvarei T, Popa V (2005) Nutrition and animal feeding. ALFA Publishing House, Iaşi, Romania

Hura C (2006) Laboratory guide—analysis methods for food products. Cermi Publishing House, Iaşi, Romania

Kannan K, Tanabe S, Ramesh A, Subramanian A, Tatsukawa R (1992) Persistent Organochlorine residues in foodstuffs from India and their implication on human dietary exposure. J Agri Food Chem 40:518–524.

Koerner P (2001) Clorinated pesticides. J Chromatogr 205(1):11–16

Legis—Romanian legislation

Singh R, Kumari B, Madan VK, Kumar R, Kathpal TS (1997) Monitoring of HCH residues in animal feed. Indian J Anim Sci 676:250–252.

Škrbič B, Predojevič Z (2008) Levels of Organochlorine Pesticides in Crops and Related Products From Vojvodina, Serbia: Estimated Dietary Intake. Environ Contam Toxicology 54(4):628–623

SR EN 12393-2 (2003) Non-fatty foods—multiresidue methods for the gas chromatographic determination of pesticides residues—part 2: Methods for extraction and clean-up

SR EN 12393-3 (2003) Non-fatty foods—multiresidue methods for the gas chromatographic determination of pesticides residues—part 3: Determination and confirmatory tests

Tadeo JL (ed) (2008) Analysis of pesticides in food and environmental samples. CRC Press, Taylor & Frances Group Boca Raton, London, New York

Wang G, Yao X-P, Cong Z-Y, Yan X-L, Kang S-C, Zhang Y (2007) Distribution of persistent organic pollutants in soil and grasses around Mt. Qomolangma, China. Arch Environ Contam Toxicol 52(2):153–162

Chapter 9
Size Exclusion Chromatography a Useful Technique For Speciation Analysis of Polydimethylsiloxanes

Krystyna Mojsiewicz-Pieńkowska

Abstract Size exclusion chromatography (SEC) is one of the most popular methods for separation and molecular characterization of natural or synthetic macromolecules mixtures. This chapter describes the retention mechanisms for SEC as ideal size-exclusion, which separates molecules primarily on the basis of their hydrodynamic volume, and non-ideal size-exclusion, which depends on entropic and enthalpic retention mechanisms and applies when adsorption occurs. Furthermore, recent applications of SEC in the biomedical and pharmaceutical sciences are shown. Finally, the use of SEC as a technique for speciation analysis of polydimethylsiloxanes (PDMS) is presented. PDMS belong to polysiloxanes, with the common name silicones, and are used in a wide variety of applications, such as in the medical (e.g. implants), pharmaceutical (e.g. the active pharmaceutical substances dimeticone and simeticone) and food (e.g. the food additive E 900) areas. Speciation analysis of this polymer is very important, because the degree of polymerization and the particle size, (or molecular weight) have experimental impacts on the toxicity, absorption and migration in living organisms.

9.1 Introduction to SEC

Size exclusion chromatography (SEC) is the most common and efficient method to obtain information about the molecular weight and molecular weight distribution of macromolecules. The use of this technique was initiated around 1950, but the primary stimulus for its rapid development was the work undertaken in the Swedish chromatograhic school, which resulted in the fundamental publication by Porath and Flodin in 1959 about the separation of water-soluble macromolecular substances with varying degrees of cross-linked dextran (Gooding and Regnier 2002; Porath

K. Mojsiewicz-Pieńkowska (✉)
Faculty of Pharmacy, Department of Physical Chemistry, Medical University of Gdańsk,
Al. Gen. Hallera 107 Gdańsk 80-416, Poland
e-mail: kpienk@gumed.edu.pl

Inamuddin, A. Mohammad (eds.), *Green Chromatographic Techniques*,
DOI 10.1007/978-94-007-7735-4_9, © Springer Science+Business Media Dordrecht 2014

and Flodin 1959). In the case of macromolecules with hydrophilic properties, the term gel filtration chromatography (GFC) has been used by various authors. The first attempt to separate synthetic polymers in a non-aqueous medium was made by Brewer and Vaughan (Gooding and Regnier 2002; Berek et al. 1983). However, the turning point in the method development was the work by Moor, who in 1964 synthesized a series of gels with selective pore dimensions (Moore 1964). Using those gels (styrene/divinylbenzene crosslinked copolymers) Moor successfully separated polystyrenes according to their molecular weights. Moor named the method gel permeation chromatography (GPC) which is still a frequently used analytical term, particularly for macromolecules with lipophilic properties.

However, when reviewing the published literature, it becomes clear that the nomenclature used in this field is inconsistent, which may cause confusion. Currently, three terms are used: SEC, GPC, and GFC. In practice, there are two types of size exclusion chromatographic techniques. One is (GPC), which uses a hydrophobic column packing material and a non-organic mobile phase (e.g. for studies of synthetic polymers). For this case, the term SEC is also used synonymously. The second type is (GFC), which uses a hydrophilic packing material and an aqueous mobile phase (e.g. for studies of molecules soluble in water, such as polysaccharides and proteins). The International Union of Pure and Applied Chemistry (IUPAC) recommends the use of two terms: SEC and GPC. According to the definition recommended by IUPAC, size exclusion chromatography is *"A separation technique in which separation mainly according to the hydrodynamic volume of the molecules or particles takes place in a porous non-adsorbing material with pores of approximately the same size as the effective dimensions in solution of the molecules to be separated"*, and gel permeation chromatography is *"Separation based mainly upon exclusion effects, such as differences in molecular size and/or shape (e.g. molecular-sieve chromatography) or in charge (e.g. ion-exclusion chromatography). The term gel-permeation chromatography is widely used for the process when the stationary phase is a gel"* (Templeton et al. 2000; IUPAC Gold Book 2006). IUPAC does not recommend the use of the term gel filtration chromatography (Templeton et al. 2000). It should be noted, however, that for SEC, GPC, and GFC the separation mechanism is the same, hence, the nomenclature should be uniform. Therefore, based on the separation mechanism, the term size exclusion chromatography should be uniformly used.

9.2 SEC Retention Mechanisms

The phenomenon of chromatographic separation involves different mechanisms, because the various techniques are based on different theoretical foundations. The methods are based on one of the following principles: adsorption, partition, ion-exchange, affinity, or size exclusion. All those are based on the same idea of separating the individual components of a mixture. Generally, the principle of chromatography is based on the differential migration of individual substances in a heteroge-

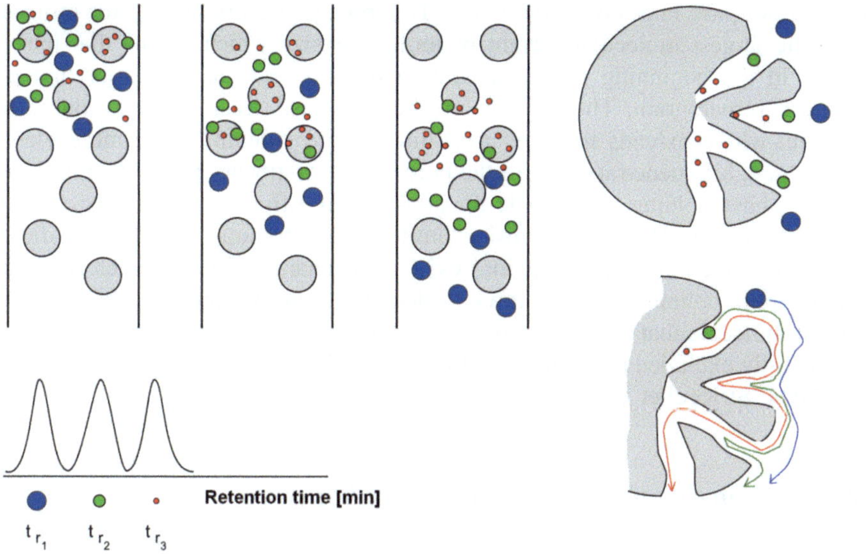

Fig. 9.1 Principal mechanism for separating molecule mixtures by size exclusion chromatography

neous system consisting of two phases: a stationary solid phase and a liquid or gas phase. However, the use of chromatographic techniques require a conscious choice, because not all techniques can be used for the selected analyte. It is therefore important to choose the most suitable technique, which depends on the physicochemical properties of the analyte, *e.g.* particle size, shape, molecular weight, volatility, hydrophilicity, lipophilicity, polarity, tendency to form hydrogen bonds, Van der Waals bonding, and possibility to form ions.

9.2.1 *Ideal Size Exclusion Mechanism*

The principal mechanism involved in separation by exclusion chromatography is not very complicated. The analytes are separated between the mobile and the stationary phase based on their size difference. In this system, the mobile phases carries some of the components in the injected sample and delivers them to the stationary phase. The latter is characterized by a porous structure, which in turn determines the nature of separation mechanism. In exclusion chromatography separation is based on differences in flow delays for molecules with different hydrodynamic radius, which in practice means molecular weight. This is a result of molecules diffusing into the solvent trapped in the pores of the stationary phase (Fig. 9.1). Particles that are larger than the pore size (with the largest hydrodynamic radius) will not diffuse to the pores, but pass in the space between them. This condition is called exclusion. Total exclusion (exclusion limit) occurs for particles that do not pass into any of the pores. This process occurs rapidly for molecules with the largest molecular weight or molecular size which means that they are eluted in the beginning of the separation process and constitute the first peak in the chromatogram. The smaller are the particles, larger is the available pore volume, which extends the flow path through the column. If the molecules are small enough to penetrate into all the pores of the stationary phase, then the entire mobile phase volume is available for non-limited diffusion, and this is called the penetration limit. Between exclusion limit and penetration limit, the specific size exclusion region (separation region) exists. It is characteristic for each SEC chromatographic column, which allows to determine the molecular weight from the largest molecule that is fully included in the pore-volume to the smallest molecule that is fully excluded (Gooding and Regnier 2002; Berek et al. 1983; Berek 2010; Wu 2004; Striegel et al. 2009).

9.2.2 Non-Ideal Size Exclusion Mechanism

Even though size exclusion is the main retention mechanism in SEC, it is not the only one. Generally, the goal is to eliminate sorption interactions between the surfaces of the column packing particles and the separated substances, but this is not always possible. Then the separation mechanism becomes thermodynamically more complicated. Sorption interactions are primarily associated with adsorption phenomena between the analyte and stationary phase packing. This is defined as the mechanism of adsorption and surface interaction. In this mechanism, the adsorbate-adsorbent interaction relies on Van der Waals forces or formation of hydrogen bonds. The Van der Waals interaction is important for the creation of the so-called instantaneous dipole moment, or London dispersion forces.

Many authors have drawn attention to the phenomenon of adsorption, and thermodynamic imbalance by entropy (ΔS) and enthalpy (ΔH) changes during the separation (Berek 2004a, 2010; Podzimek 2011; Russ and Berek 2007). Adsorption is a spontaneous process and is associated with thermodynamic changes (thermodynamic functions). To describe the effects associated with the adsorption energy, several thermodynamic terms have to be used, e.g. free enthalpy of adsorption, adsorption entropy, and adsorption enthalpy. Therefore, in size exclusion chromatography the entropic and enthalpic mechanisms exist. The entropic retention mechanism relates to the size exclusion that results from the variation in particle size of molecules. Changes in entropy are caused by concentration gradients, the flow within the column, and the diffusion process. This results in partial or full exclusion of macromolecules from the pores or from the outer surface of the sorbent particles. This entropic mechanism is recognized as the basic retention mechanism in SEC. If there are interactions between the analyte and the stationary phase, or between the mobile and stationary phase, this adds an additional enthalpic retention mechanism (Berek 2004b, 2010; Podzimek 2011; Russ and Berek 2007; Trathnigg 2000).

Additional retention mechanisms can add undesirable effect, e.g.: lower selectivity of column and the erroneous molar weight estimation. Some examples to cite are

- peak broadening due to inappropriate choice of column;
- diffusion effects;
- inadequate separation of peaks in the mixture;
- illogical elution order of the components in the mixture (polymer molecules that strongly bind to the sorbent will leave the column later regardless of their size).

To avoid damaging adsorption phenomena, one can try to change either the elution strength or the flow rate of the mobile phase, or the entire column.

9.3 The Stationary Phase in SEC

In size exclusion chromatography, both classical particulate and monolithic packing materials are used. Monoliths are novel materials for this chromatographic technique. The advantage of monolithic stationary phases is that they contain a significant amount of small pores within their pore size distribution (Wu 2004; Trathnigg 2000; Ali et al. 2009). In each column filled with porous particles, intergranular pores are formed, through which the mobile phase flows and fills a large number of open micropores inside the grains. In monolithic columns, the filling contains both large-sized mesopores and micropores within the "solid matrix", though the liquid hardly flows through the micropores. The pores within the particulate column packing material, or the micropores in the monolithic columns, always have different diameters to some extent. In size exclusion chromatography columns, the mean value and the characteristics of the pore diameter distribution are selected based on the hydrodynamic radius of the separation particles. This is related to the molecular weight distribution of the molecules in sample. The stationary phases, used in SEC (Table. 9.1), are divided into the following types (Trathnigg 2000; Ali et al. 2009; Grimes et al. 2007; Li et al. 2009; Kale et al. 2007; Wu 1999; Walsh et al. 2010; Gaborieau and Castignolles 2011; Kostanski et al. 2004):

1. organic and non-organic—based on the chemical structure;
2. wet swollen and hard—relates to the interactions with the mobile phase;
3. lipophilic and hydrophilic—based on the properties of the substances to be separated.

9.4 The Mobile Phase in SEC

A prerequisite for size exclusion chromatography is that the mobile phase (the eluent) be a good solvent for the analyte. The mobile phase must be selected to avoid destroying the column packing material, and the elution strength should be

Table 9.1 Classification and characterization of stationary phases used in size exclusion chromatography (SEC)

Types of stationary phases for SEC	Feature	Name	Examples	Characteristic
Classical packing material	Chemical structure	Organic	Dextran based, e.g. Sephadex LH-20, Sephacryl	Physically stable
			Agarose based, e.g. Sepharos, Biogel A, Ultragel A, Work Beads 17 SEC, Superose 12 HR	Wide pore distribution
			Acrylate, methacrylate, e.g. Biogel P, Ultragel ACA	
			Polystyrene—divinylbenzene, e.g. PS-DVB, Lichrogel PS 1-4000, TSK gel G1000-G7000	
			Branched polymers	
		non-organic	Silica, e.g. Lichrospher SI 100-4000, TSK-GEL SW 2000-4000	Deactivated surface sorption
			Porous glass, e.g. CPG 40-3000	High physical stability
			Zirconium oxide aluminum oxide	
	Interaction with mobile phase	Wet swollen	Polymethacrylate, e.g. Toyopearl HW-40	Requires preparation before use—swelling
			Hydroxypropylated dextran, e.g. Sephadex LH-20	High mass transfer rate
				Low back pressure
		Hard	Methacrylate co-polymer, e.g. Toyopearl HW-40	Deactivated surface sorption (blocks strongly polar groups, e.g. −OH and =O, on the surface)
			Poly(styrene-co-divinylbenzene, e.g. TSK gel $H_{HR}GMH_{HR}$	
			Silica	
			Porous glass, e.g. BioGlas	
			Zirconium oxide	
	Properties of the substances forseparation	Lipophilic	Dextran based, e.g. Sephadex LH-20	High resolution separation
			Poly(styrene-co-divinylbenzene, e.g. TSK gel $H_{HR}GMH_{HR}$	Excellent reproducibility, excellent durability
		Hydrophilic	Silica	Deactivated surface sorption (blocks the strongly polar −OH group)
			Porous glass, e.g. BioGlas	
			Methacrylate, e.g. Biogel	
Monolithic packing material	Chemical structure	Organic	Polymeric monolith, e.g. methacrylate – based, polystyrene-co-divinylbenzene	Highly porous
		Non-organic	Silica monolith, e.g. Chromolith	Highly homogeneous
				More suitable for high resolution
				Exhibit a distinct bimodal pore volume distribution with macropores and mesopores—fast separations at very low column back pressures
				High column performance

sufficient to eliminate adsorption of the polymer molecules to the surfaces of the packing material (Gooding and Regnier 2002; Berek et al. 1983; Berek 2010; Wu 2004; Trathnigg 2000). Selection of the mobile phase determines what would be an appropriate stationary phase. Generally, the mobile phases can be divided into two basic groups, depending on the separation conditions:

a. Non-aqueous, i.e. lipophilic conditions: tetrahydrofurane, dioxane, tetrachloroethylene, chlorobenzene, dichlorobenzene, toluene, xylene etc. As the stationary phase, copolymers of styrene and divinylbenzene, or polyesters, are used because these are resistant to organic solvents. This system is used to determine the molecular weight, and weight distributions, of polymers with low and medium polarity that are soluble in non-polar solvents (e.g. polystyrene in toluene), lipids, phospholipids, waxes, and other non-polar substances of a natural origin.
b. Aqueous solutions, or strongly polar non-aqueous (hydrophilic) eluents, such as dimethylformamide, methanol, acetonitrile, and mixtures of these with water, as well as aqueous solutions of salts, acids and bases.

As the stationary phase, polar materials are used, such as polydextranes and other polysugars, polycarbonate, porous glass, or silanized silica gel. Such systems are used for separation and characterization of molecular weight distributions of polar polymers (e.g. polyethylene glycol in water), particularly biopolymers (polysaccharides, proteins, nucleotides).

9.5 Analytical Problems

There are many potential problems that can appear during the analysis, and the most important are:

- Adsorption interactions between the separated polymers and the stationary phase;
- Swelling and contraction of the particulate filling material in the columns with different eluents; a poorly chosen mobile phase or eluent can rapidly change the column and even permanently destroy it. It is important to observe the general rule that in columns used for separation of synthetic polymers with low and medium polarity, highly polar liquids such as water, methanol, ethanol, isopropanol, and acetonitrile should not be used. Otherwise an irreversible contraction of the particles in the filled columns may occur, with a subsequent loss of both the column selectivity and efficiency. Similar problems can occur in columns filled with soft and semi rigid hydrophilic materials, which are made after the material has swollen in an aqueous solution. In this case, hexane or toluene, can destroy the pore structure of the particulate filling, and acetone or THF may cause excessive swelling of the packing material, or even dissolve it. In the latter case, the content of the column may flow into the detector and damage it. The following column problems arised.
- Limited compressibility of the packing material (even hard);

- Poor pressure resistance of the packing material;
- Free diffusion of macromolecules, and the need to reduce the eluent flow rate to avoid excessive dilution of the peaks;
- Competition between molecules for diffusion into the pores;
- Pore clogging by some macropolymers;
- Polycondensation of macromolecules under high pressure, causing an increase in the molecular weight of the sample.

9.6 Methods for Column Calibration

To determine the correct molecular weight and molecular weight distribution, the calibration of the selected column SEC is needed. There are three methods of calibration, and the choice for a particular analysis is dependent on the capabilities and quality requirements. The following methods are available for calibrating the column (Berek et al. 1983; Berek 2010; Trathnigg 2000; Kostanski et al. 2004):

a. Calibration with Narrow Standards—the elution volumes, or retention times, for a series of standards with a narrow molar mass distribution (MMD) is determined. As an approximation, a linear relation between the logarithm of the molecular weight (Log M) and the elution volume (Ve) or retention time (tr) is assumed (Fig. 9.2). Despite the fact that the function is often sigmoidal (Fig. 9.2), a linear regression is used for a specific size exclusion region. Therefore, the calibration function is quite simple in this case. Calibration with Narrow Standards is a method with high accuracy.
b. Calibration with Broad Standards—In this case the standards are characterized by considerable degree of polydispersity and therefore the molar mass distribution (MMD) must be determined for the calibration. Calibration with broad standards is a method with lower accuracy.
c. Universal Calibration—This is based on the relationship described by the Mark-Houwink equation ($[\eta]=K\cdot M^a$), and knowledge of the constants K and a for given polymers. Based on experimental results, it has been shown that the product of $[\eta]$ and Mw (where η is the intrinsic viscosity, and Mw the average molecular weight) can be used as a universal parameter. This is justified because the product $[\eta]$ M is proportional to the hydrodynamic volume of the polymer.

For calibration of the chromatographic column, it is best to use a standard with the same chemical structure as the analyte. However, this is not always possible because all polymers are not available as SEC standards. Since there is often only one primary separation mechanism (ideal size exclusion), it is possible to use other polymers than those tested as standards. However, it is necessary to check the accuracy of the molecular weight determination to make sure that the results are correct. If there is an ideal size exclusion mechanism, polymers with different chemical identity have the same relationship between log M and the retention time (tr). This enables a broad application of size exclusion chromatography, in which the

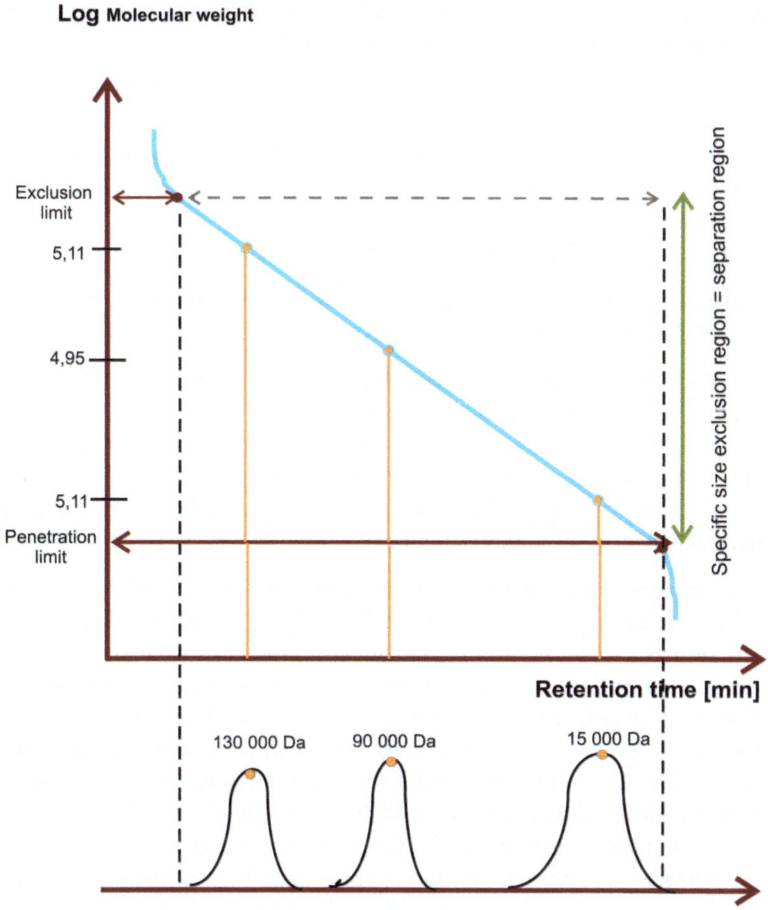

Fig. 9.2 Column calibration curve with Narrow Standards in size exclusion chromatography

analyzed polymer is not the same as the polymer used for calibration of the column (Temyanko et al. 2001).

In studies of polymers with low and medium polarity, it is common to use polystyrene standards, with the ratio maximum to minimum molecular weight, M_{max}/M_{min}, in the range of 1.05 to 1.20. In most studies of polar polymers, polyethylene glycol is used. polystyrene and polyethylene glycol are considered as universal standards (Kostanski et al. 2004; Temyanko et al. 2001; Garcia-Lopera et al. 2003).

9.7 Applications of SEC Biomedical and Pharmaceutical

Currently, size exclusion chromatography has been an indispensable method for studies of macromolecular compounds, both oligomers and polymers. The following applications are worth to mention. (Gooding and Regnier 2002; Porath and Flodin 1959; Berek et al. 1983; Moore 1964; Berek 2010; Wu 2004; Striegel et al. 2009; Trathnigg 2000; Gnanou and Fontanille 2008):

i) Separation of mixtures, identification and quantitative studies of macromolecular analytes,
ii) Separation of low molecular weight substances from oligomers or polymers,
iii) Purification of small-size molecules from large-size molecules,
iv) Determination of molecular weights,
v) determination of molecular weight distributions (polydispersity).

SEC methods have been used to separate, identify and quantify the following types of polymers (Berek et al. 1983; Berek 2010; Gnanou and Fontanille 2008):

a. Natural polymers from plants, (cellulose, starch and natural rubber), and from animal (proteins such as wool and silk).
b. Semi synthetic polymers obtained by chemical modification of natural polymers such as cellulose esters (nitrocellulose, cellulose acetate, etc.);
c. Synthetic polymers prepared in laboratories by polymerization of monomer molecules such as polysiloxanes (silicones).

Size exclusion chromatography is used in many fields, and in the last decade it has received an increasing interest from the biomedical and pharmaceutical research fields. Examples can be found in biomedical research on systems biology, which integrates data in order to define a number of phenomena occurring in the organisms using mathematical models (Hertog et al. 2011). Others are some of the dynamically developing technological platforms such as genomics, transcriptomics, proteomics, lipidomics, and metabolomics. Technological advances, both in the platforms and in the development of analytical methods, can help to develop an understanding of complex biological systems in the context of the functions, parameter interdependence, interactions, and irregularities that are main causes of several diseases. Examples of applications of SEC in technological platforms in certain fields of systems biology are listed in Table 9.2, and the types kind of information that can be obtained for individual systems biology platforms using size exclusion chromatography are shown in Fig. 9.3.

In recent years, many publications have appeared on the use of size exclusion chromatography in pharmaceutical sciences. This applies both to development processes for new medicinal products, pharmaceutical availability of formulations (e.g. from tablets), and the bioavailability of active pharmaceutical ingredients (API). An important field is the control of medicines and medical devices, to meet the basic quality and safety requirements for such products. Furthermore, size exclusion chromatography constitutes a method for continuous monitoring of all the development stages of nanoparticulate drug delivery systems, thereby ensuring the quality of both the starting materials and the final products. The primary properties that can be monitored with this technique include particle sizes and molecular weights ((Kar 2005; Mao et al. 2004; Williams et al. 2002; Shan-Ying et al. 2003; Otto et al. 2007; Mojsiewicz-Pieńkowska 2012).

Table 9.2 Examples of size exclusion chromatography applications in some systems biology technology platforms

Technology platform	Platform purpose	Examples of size exclusion chromatography applications	References
Genomics	Study of organisms DNA[a] sequence determination, including intra-organismal cell specific variation (i.e. telomere length variation)	The unknown histone H2A.Z variant was indentified in human cell of brain and a comprehensive characterization of its nucleosomal properties was provided	(Kato and Shigeru 2004; Bonisch et al. 2012)
Transcriptomics	Determination of entire sets of RNA[b] molecules, including mRNA, rRNA, tRNA, and other non-coding RNA produced in one cell, or a population of cells	The incorporation of transcriptomic approaches into ecological and evolutionary studies helps further exploring how natural populations respond to environmental change and anthropogenic pressures	(Kato and Shigeru 2004; Bonisch et al. 2012; Pujolar et al. 2012)
Proteomics	Science dealing with the study of proteins and the relationship between protein structures and their biological functions	High-molecular-weight -amyloid oligomers were determined in cerebrospinal fluid of Alzheimer patients analysis of protein-extracts obtained from bacterial cultures grown Report showing 52 proteins as candidate serological ovarian cancer biomarkers	(Fukumoto et al. 2010; Lecchi et al. 2003; Kuk et al. 2008)
Lipidomics	Study of pathways and networks of cellular lipids in biological systems	Determination of lipids as abundant constituents of both the vascular plaque and lipoproteins, which play a pivotal role in atherosclerosis	(Ekroos et al. 2010)
Glycomics	Describes the complete set of glycans and glycoconjugates, which the cells produce under specified conditions of time, space, and environment	Functional glycomic analysis of human milk glycans reveals presence of virus receptors and embryonic stem cell biomarkers	(Yu et al. 2012)
Metabolomics	Organism, tissue, or cell level measurements of all small molecules known as metabolites	Study of metabolomics in diabetes research by identification of a series of known and novel, deregulated metabolites	(Suhre et al. 2010)

[a] deoxyribonucleic acid (DNA)
[b] ribonucleic acid (RNA)

9.7.1 SEC as a Useful Technique for Linear Polydimethylsiloxanes Speciation Analysis

Polydimethylsiloxanes (PDMS) belong to the polysiloxanes (silicones) (Fig. 9.4) and are currently used in a wide variety of applications. Some of the most important applications include as medicinal products (pharmaceutical ingredients dimethi-

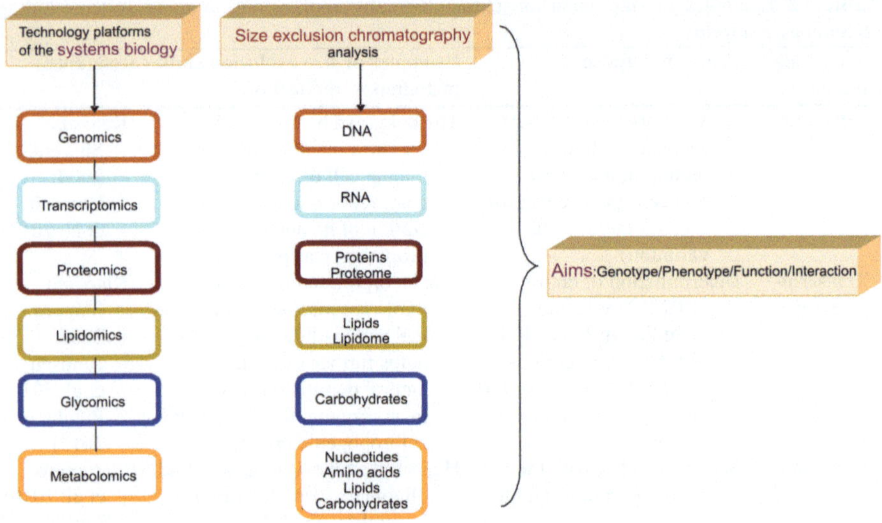

Fig. 9.3 Application of size exclusion chromatography in technology platforms within the systems biology field

cone and simethicone, and as excipients), the adhesive patches designed to deliver therapeutic substances through the skin such as the patch for delivering API dermally), medical devices (e.g. implants, dentures, for ocular endotamponade to treat complicated retinal detachments, silanized syringes, tubes, and catheters) and food (as the functional additive E-900) (Mojsiewicz-Pieńkowska 2012; Curtis and Colas 2004; Subedi et al. 2010; Sweetman 2009; Moore et al. 2002; Torrado et al. 1999).

Polydimethylsiloxanes (PDMS) belong to the organosilicon polymers. The structure is basically a backbone with repeating silicon-oxygen atoms (–Si-O–) and methyl substituents (Fig. 9.4). The molecules are formed by chain reactions of small molecules called monomers [–Si(CH_2)-O–]. Figure 9.4 shows the general structure of linear PDMS, where the degree of polymerization (n is the number of repetitive units in the molecule) determines the molecular weight and the viscosity. Hence, the PDMS group can be divided into low molecule, middle molecule and high molecule weight PDMS. Due to the specifics of the polymerisation reaction, it results into a product that must be treated as a mixture of polymers, including oligomers, with chains of variable lengths and consequently variable molecular weights. Hence, PDMS are not individual chemical species, which is an important issue from the toxicological point of view. Current knowledge confirms that the structure and size of the molecules affect the absorption and migration in animal and human bodies, as well as the toxicity. The issue of silicones safety is all the more important as not only adults and children are in contact with them, but also infants. In the case of food, the oral use of PDMS is limited to a polymerization degree of $n=90$–410, corresponding to molecular weights of 6,800–30,000 Da, and viscosities of 100–1,500 cSt. For medicinal products, the range is $n=50$–400, corresponding to molecular weights of 3,500–30,000 Da, and viscosities of 50–1,300 cSt. Speciation analyses, and the use

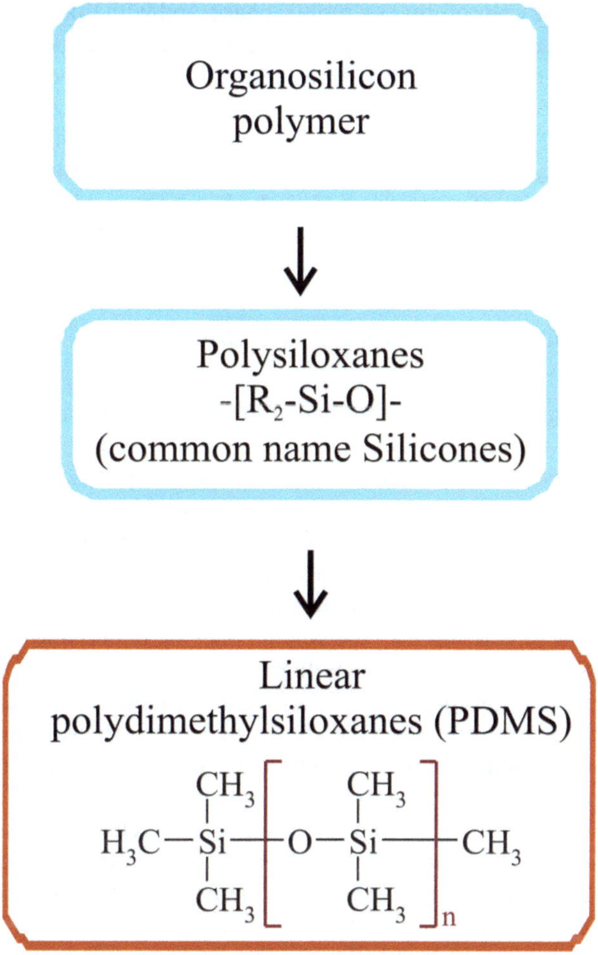

Fig. 9.4 Chemical classification and structure of polydimethylsiloxanes

of specific analytical methods, are particularly important when considering the toxicity of some of the PDMS molecules. The size exclusion chromatography satisfies the condition for speciation analysis, and is a specific method for PDMS. A speciation analysis provides reliable information about an individual (specific) chemical species in a sample contained in *e.g.* a pharmaceutical product.

9.8 Methodology for Linear Polydimethylsiloxanes Speciation Analysis

Combining size exclusion chromatography with an evaporative light scattering detector (SEC-ELSD) provides a method for speciation analysis of linear polydimethylsiloxanes that differ in chain length, molecular weight and viscosity. It is possible to use chloroform as a mobile phase, and with this a high repeatability (precision) of the retention time and the peak surface areas can be obtained. This method has been proven specific for PDMS, because PDMS that differed in molecular weight and viscosity were accurately separated (Mojsiewicz-Pieńkowska 2008, 2010, 2011, 2012). So far, problems with speciation analysis of linear polydimethylsiloxanes are not widely known. A review of the literature published since 1970 indicates that the majority of currently used methods are not specific enough for PDMS speciation in pharmaceutical formulations and foodstuffs (Mojsiewicz-Pieńkowska 2012).

9.8.1 Mobile Phase Selection

Contrary to other chromatographic techniques, in size exclusion chromatography the selection of the separation conditions starts by selecting the optimum mobile phase. The type of mobile phase determines the choice of the most suitable stationary phase, and hence this stage is crucial in defining the analytical procedure. The main criterion for the selection of the mobile phase in the separation, identification and quantitative research of PDMS is a very good solubility of these polymers. When selecting the mobile phase also other requirements have to be met, such as a low interaction with the column packing material to ensure stability of the column, and chemical inertness with respect to the macromolecules, as well as noise separation during formation of the signal from the laser light scattering detector (ELSD).

Solubility tests have shown that linear PDMS are soluble in chloroform, tetrahydrofuran, toluene, and methyl isobutyl ketone, and do not dissolve in water, alcohols (methyl, ethyl, isopropyl) and acetonitrile. In view of the appropriate boiling point (61.22 °C), chloroform is often chosen as the mobile phase. The boiling point is important, as the solvent must evaporate before the detection of the analyte. Evaporation of chloroform in the detector does not cause any formation of noise, and the detector signal is thus characterized by a high intensity, which gives results that are reproducible compared to other organic solvents.

9.8.2 Stationary Phase Selection

In the case of linear PDMS, the choice of the stationary phase is based on the fact that PDMS are soluble in hydrophobic solvents and have significantly different molecular weights when occur in oral medicines and food (in food: $n=90–410$, corresponding to molecular weights of 6,800–30,000 Da and viscosities of 100–1,500 cSt; in medicinal products: $n=50–400$, corresponding to molecular weights of 3,500–30,000 Da and viscosities of 50–1,300 cSt), as well as in polymers not permitted in those products ((Mojsiewicz-Pieńkowska 2012). For these reasons, the column TSK-GEL-M $H_{HR}GMH_{HR}$ filled with polystyrene-divinylbenzene, bed particle size 5 μm, has been selected for general use. The manufacturer guaranteed specific size exclusion region (separation region) range from 500–1,000,000 Da.

Table 9.3 Solvents that are compatible with the TSK-GEL $H_{HR}GMH_{HR}$-M column packing material

Toluene	Benzene	Dioxane
Chloroform	Dichloromethane	m-cresol/chloroform
n-heksane	Dimethylsulfoxide	trichlorobenzene
Carbon tetrachloride	Cyklohexane	Pyridine
Methanol/chloroform	Xylene	Ethyl acetate
Trichloroethane	Dichloroethane	Acetone

9.8.3 Column Conditions

Any column that is used in size exclusion chromatography is filled by the manufacturer with a polar or non-polar solvent. In the case of the column TSK-GEL $H_{HR}GMH_{HR}$-M polystyrene-divinylbenzene, it saturated with tetrahydrofuran. Thus, the column can not be used directly for studies where the mobile phase is chloroform. The only way to avoid damaging the column packing material and get optimal column separation, the use of the procedure of change the solvents is necessary. First it has to be checked that the solvents are compatible with the column, and will not cause loss of the pore size distribution and stability of the stationary phase. Table 9.3 lists options (approved by the manufacturer) for replacement of tetrahydrofuran, which is the original solvent conditioning the column, with a different solvent.

Each solvent listed in Table 9.3 may be used as the mobile phase for separation of a mixture of analytes in a column of this type, provided that it will not contain admixtures of water, which reduces the efficiency of the column and can affect, or even damage, the packing material. The solvent exchange process can be carried out as outlined below. First, mixtures of tetrahydrofuran and chloroform are prepared with the following composition:

- a mixture of 75 % tetrahydrofuran and 25 % chloroform;
- a mixture of 50 % tetrahydrofuran and 50 % chloroform;
- a mixture of 25 % tetrahydrofuran and 75 % chloroform.

Each mixture is subsequently passed through the column at a flow rate of 0.3 ml/min for 90 min, starting with the ratio of 3:1 of tetrahydrofuran and chloroform, and gradually increasing the chloroform content up to 100 % in the final solvent. This procedure allows for complete saturation of the column with chloroform.

9.8.4 Column Calibration

Based on calibration curves presented by various authors, it is clear that the relationship between the logarithm of the molecular weight (log M), time (t_r) and retention volume (V_r) is not always linear. Furthermore, the shapes of the curves for columns operating in various ranges are very different. We can therefore conclude

Table 9.4 Characteristics of polystyrene standards

Molecular weight at peak—M_{SEC} [Da]	log M_{SEC}	Average molecular weight—M_w [Da]	Number average molecular weight—M_n [Da]	Polydispersity factor—M_w/M_n
2,570,000	6.4099	2,530,000	2,420,000	1.05
1,090,000	6.0374	1,070,000	1,010,000	1.06
702,000	5.8463	682,000	651,000	1.05
250,000	5.3979	248,000	236,000	1.05
130,000	5.1139	125,000	120,000	1.04
67,500	4.8293	65,000	64,000	1.02
34,700	4.5403	34,000	32,700	1.04
17,800	4.2504	17,500	17,000	1.03
8,400	3.9243	8,100	7,800	1.04
3,420	3.5340	3,470	3,280	1.06
1,620	3.2095	1,560	1,500	1.04
376	2.5752	484	435	1.11

that the calibration curves are affected by many factors. These include, inter alia, the nature of the mobile phase, and the grain and pore sizes of the column filling. Therefore, it is necessary to determine the calibration curve for a particular stationary phase using a certified standard with fixed polydispersity.

In PDMS studies, certified polystyrene standards (Sigma-Aldrich) are used for column calibration. The molecular characteristics of polystyrene standards are presented in Table 9.4. All of them have a polydispersity factor between 1.02 and 1.11, which means that they are characterized by a narrow molecular weight distribution. A chromatogram of a polystyrene standard mixture separated in a mobile phase at a speed of 0.3 ml/min, and the column calibration curve, is shown in Fig. 9.5. The resulting relationship between the logarithm of the molecular weight at the peak, M_{SEC}, for the polystyrene standards, and the retention time t_r was a linear function: $y = -0.2526 \times 10.613$, with an $R^2 = 0.9983$. The TSK-GEL column $H_{HR}GMH_{HR}$-M has a wide operating range, which is 2,570, 000–376 Da.

9.8.5 Separation of Polydimethylsiloxanes

The selected column should efficiently separate a mixture of linear PDMS compounds with different degree of polymerization, (or molecular weights) and viscosities. This separation is based on an exclusion mechanism, where the PDMS with the smallest molecular weight has the longest retention time, because it penetrates deeply into the pores of the column material. The SEC-ELSD method is useful for quality control of medicines containing dimeticone or simeticone, and of food. On the basis of the retention time of the peak, the molecular weight of PDMS can be determined. Such data are important to determine whether the polymer is in compliance with recommendations (e.g. European Pharmacopoeia—EP, United States

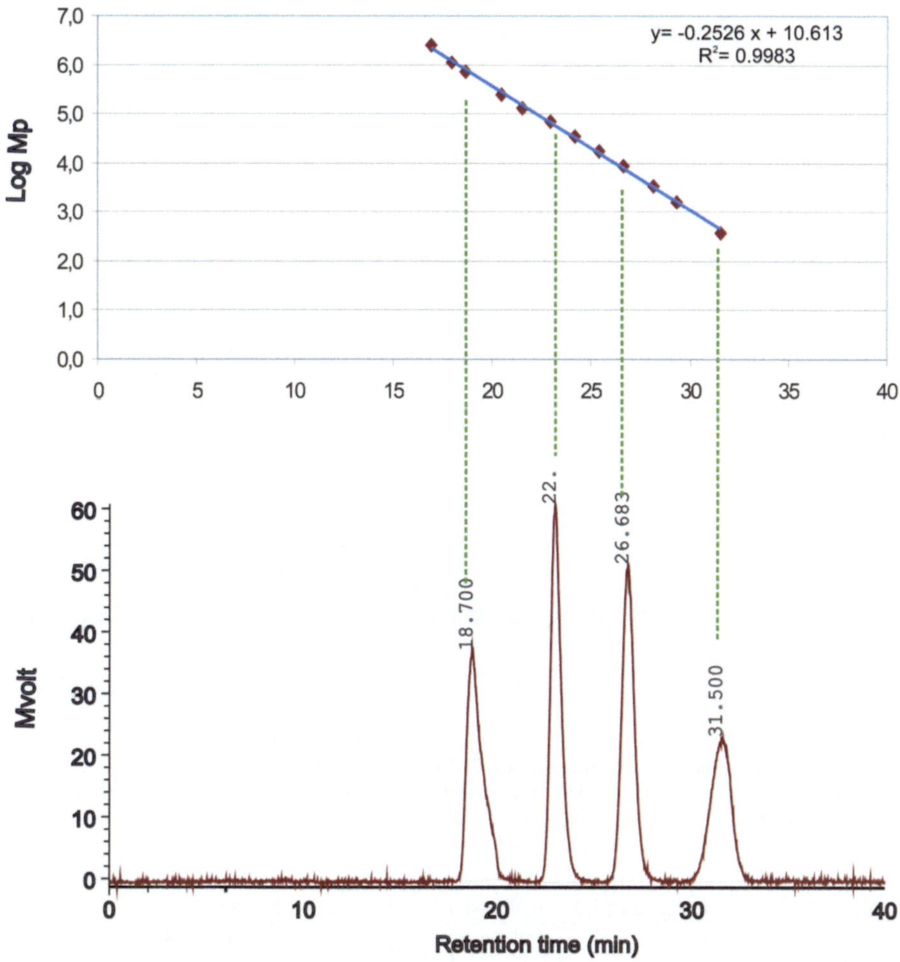

Fig. 9.5 Chromatogram for separate mixtures of narrow polystyrene standards (702,000–376 Da)

Pharmacopeia—USP, US Food and Drug Administration—FDA, World Health Organization—WHO, The Joint FAO/WHO Expert Committee on Food Additives—JECFA). With the analytical procedure described, the best resolution was obtained for PDMS with a viscosity of 50 cSt peak, whereas the peaks for PDMS with 350 and 60 000 cSt had a partial overlap. However, this did not prevent an accurate retention time determination, as confirmed by Fig. 9.6. For more detailed assessment of the separation at a flow-rate of 0,3 ml/min, a resolution column was calculated (Table 9.5).

The SEC-ELSD method is also useful to assess the polydispersity of PDMS. This is important as polymers can contain undesired fractions as impurities origi-

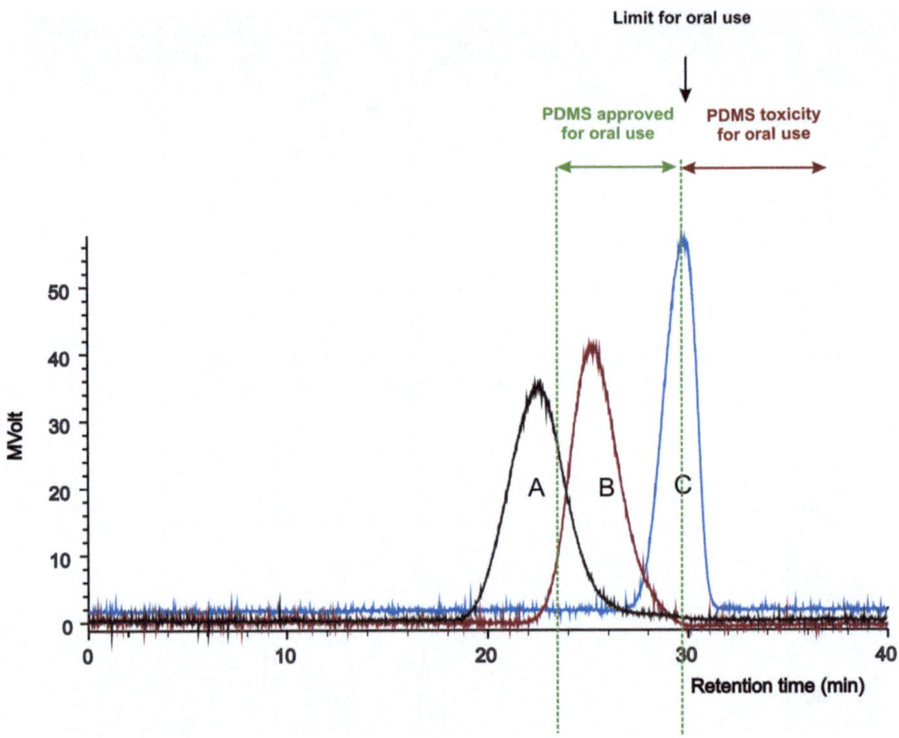

Fig. 9.6 Superimposed chromatograms of PDMS with different molecular sizes: *C* 3500 Da, 50 cSt; *B* 13,600 Da, 350 cSt; *A* 116,500 Da, 60 000 cSt; and the range limits for oral use, from 3500 Da, 50 cSt to 30 000 Da, 1,500 cSt

Table 9.5 Peak resolution for PDMS with viscosities 50, 350 and 60,000 cSt

Retention time mean value, $n=10$ [min]			Peak-width at base (w_b) mean value, $n=10$ [min]			Peak resolution (R_s)		
Viscosity of PDMS [cSt]			Viscosity of PDMS [cSt]			Viscosity range for PDMS [cSt]		
50	350	60,000	50	350	60,000	50–350	350–60,000	50–60,000
29.83	25.37	22.54	3.0	5.0	5.5	1.11	0.54	1.72

nating from the synthesis. Figure 9.7 shows the chromatogram of simethicone USP, which is recommended by the American Pharmacopoeia for production of drugs. By comparing the peak shape of simethicone USP with the peak for standard polystyrene, used for calibration of the column, it can be concluded that simethicone USP has a greater degree of polydispersity than the standard. However, the dominant fraction has a molecular weight of 17,800 Da, which corresponds to a viscosity of 410 cSt as declared (Dow Corning, Product Information Healthcare, Q7-2243 LVA Simethicone USP). A higher degree of polydispersity is not desirable, because the polymer may contain oligomers with an unauthorized degree of polymerization, and therefore molecular weight and viscosity.

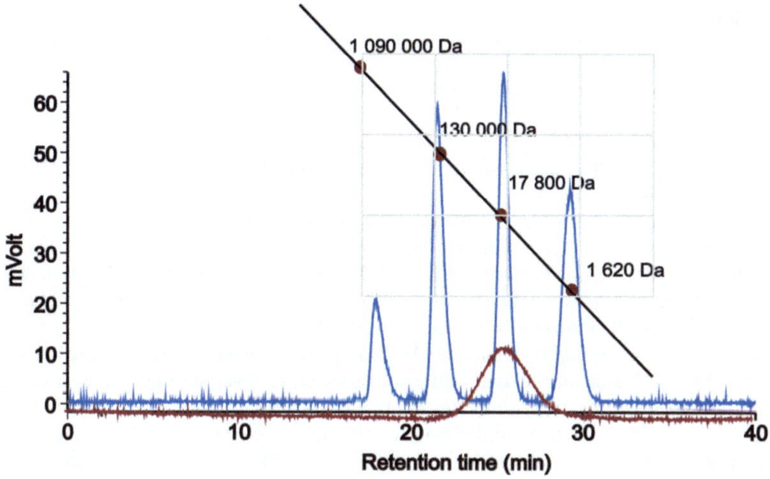

Fig. 9.7 Superimposed chromatograms of narrow polystyrene standards and the Simethicone USP product

9.9 Conclusions

All molecular characteristics of synthetic polymers exhibit a certain distribution around the mean, so that they in fact represent multicomponent mixtures of macromolecules. This is caused by the continuous nature of the polyreactions. In the case of polymer molar masses, an important and often employed parameter is the ratio of the weight average molar mass, *Mw*, to the number average molar mass, *Mn*, of a given polymer, which is called the polydispersity. Each polymer sample, however, contains one prevailing polymer with a given molecular weight, the dominant mass fraction, even though there might also be other, undesirable, polymers as contaminants. Hence, PDMS is not an individual chemical species. When considering the toxicity of polysiloxanes, one should always refer to a particular polymer.

Based on the assessment of accuracy and trueness when determining the molecular weight of PDMS with SEC (relative error = −4.41 for accuracy and relative error = −2.33 for the trueness), it can be concluded that the separation mechanism for these polymers only depend on the spatial configuration and size of the polymer molecules, and the pores of the stationary phase. Due to this lack of interaction of the analyte with the column packing material, the mechanism for PDMS separation is classified as an ideal SEC. Therefore, it can also be concluded that the suggested selection of mobile and stationary phases are optimal and that the use of narrow polystyrene standards for column calibration is suitable.

References

Ali I, Gaitonde VD, Aboul-Enein HY (2009) Monolithic Silica Stationary Phases in Liquid Chromatography. J Chrom Sci 47:432–441
Berek D (2004a) Critical assessment of the size exclusion chromatography. Polimery 5:311–386
Berek D (2004b) Enthalpic partition- assisted size exclusion chromatography: 1. Principle of method. Macromol Symp 216:145–163
Berek D (2010) Size exclusion chromatography—a blessing and a curse of science and technology of synthetic polymers. J Sep Sci 33:315–335
Berek D, Kubin M, Marcinka K, Dressler M (1983) Gel chromatography, Veda, Bratislava (publishing in Slovak); Państwowe Wydawnicztwo Naukowe, Warszawa (publishing in Polish1989)
Bonisch C, Schneider K, Punzeler S, Wiedemann SM, Bielmeier C, Bocola M, Eber HC, Kuegel W, Neumann J, Kremmer E, Leonhardt H, Mann M, Michaelis J, Schermelleh L, Hake SB (2012) H2A.Z.2.2 is an alternatively spliced histone H2A.Z variant that causes severe nucleosome destabilization. Nucleic Acids Res 40(13):5951-5964
Curtis J, Colas A (2004) Medical applications of silicones. In: Ratner BD, Hoffman AS, Schoen FJ, Lemons JE, (eds) Biomaterials science. Elsevier/Academic Press, New York, pp 698–707 (Chap. 7.19)
Dow Corning, Product Information Healthcare, Q7-2243 LVA Simethicone USP
Ekroos K, Jänis M, Tarasov K, Hurme R, Laaksonen R (2010) Lipidomics: a tool for studies of atherosclerosis. Curr Atheroscler Rep 12(4):273–281
Fukumoto H, Tokuda T, Kasai T, Ishigami N, Hidaka H, Kondo M, Allsop D, Nakagawa M (2010) High-molecular-weight (-amyloid oligomers are elevated in cerebrospinal fluid of Alzheimer patients. The FASEB J 24:2716–2726
Gaborieau M, Castignolles P (2011) Size-exclusion chromatography (SEC) of branched polymers and polysaccharides. Anal Bioanal Chem 399(4):1413–1423
Garcia-Lopera R, Irurzun I, Concepcion A, Campos A (2003) Fractal calibration in size-exclusion chromatography I. An introduction. J Chrom A 996:33–43
Gooding KM, Regnier FE (2002) Size exclusion chromatography. In: Gooding KM, Regnier FE (ed) HPLC of biological macromolecules. Chromatographic sciences, vol 87, 2nd edn. Marcel Dekker Inc, New York, pp 45–74
Gnanou Y, Fontanille M (2008) Organic and physical chemistry of polymers. Wiley, New Yersay
Grimes BA, Skudas R, Unger KK, Lubda D (2007) Pore structural characterization of monolithic silica columns by inverse size-exclusion chromatography. J Chrom A 1144:14–29
Hertog MLATM, David R, Rudell DR, Pedreschia R, Schafferc RJ, Geeraerda AH, Nicola DA, Fergusonc I (2011) Where systems biology meets postharvest. Postharvest Biol Technol 62:223–237
IUPAC the Gold Book (2006) Speciation" in IUPAC Compendium of Chemical Terminology
Kale A, Cheryan M, Robinson S, Maclaren S (2007) Surface structure of size exclusion chromatography stationary. J Microsc 227:110–117
Kar A (2005) Size exclusion chromatography. In: Pharmaceutical drug analysis, 2nd edn. New Age International Ltd., India, pp 476–484
Kato Y, Shigeru N (2004) Size exclusion chromatography of nucleic. In: Wu C (ed) Handbook of size exclusion chromatography and related techniques. Chromatographic science, vol 91, 2nd edn. Marcel Dekker Inc, New York, pp 468–486
Kostanski LK, Keller DM, Hamielec AE (2004) Size-exclusion chromatography—a review of calibration methodologies. J Biochem Biophys Methods 58:159–186
Kuk C, Kulasingam V, Gunawardana CG, Smith CR, Batruch I, Diamandis EP (2008) Mining the Ovarian Cancer ascites proteome for potential ovarian cancer biomarkers. Mol Cell Proteomics 8(4):661–669
Lecchi P, Gupte AR, Perez RE, Stockert LV, Abramson FP (2003) Size-exclusion chromatography in multidimensional separation schemes for proteome analysis. J Biochem Biophys Methods 56:141–152

Li Y, Tolley D, Lee ML (2009) Preparation of polymer monoliths that exhibit size exclusion properties for proteins and peptides. Anal Chem 81:4406–4413

Mao Y, Thompson MJ, Wang O, Tsai EW (2004) Quantitation of poloxamers in pharmaceutical formulations using size exclusion chromatography and colorimetric methods. J Pharm Biomed Anal 35:1127–1142

Mojsiewicz-Pieńkowska K (2008) Size-exclusion chromatography with evaporative light scattering detection: Method for determination of polydimethylsiloxanes. II: Application of TSK-GEL HHR GMHHR_M column to determine and separate molecular weight of linear polydimethylsiloxanes. J Chrom B 865:7–12

Mojsiewicz-Pieńkowska K (2010) Size exclusion chromatography with evaporative light scattering detection as a method for speciation analysis of polydimethylsiloxanes. I: Influence of selected factors on the signal intensity of the detector. J Pharm Biomed Anal 53(3):503–509

Mojsiewicz-Pieńkowska K (2011) Size exclusion chromatography with evaporative light scattering detection as a method for speciation analysis of polydimethylsiloxanes. II: Validation of the method for analysis of pharmaceutical formulations. J Pharm Biomed Anal 56(4):851–858

Mojsiewicz-Pieńkowska K (2012) Size exclusion chromatography with evaporative light scattering detection as a method for speciation analysis of polydimethylsiloxanes. III. Identification and determination dimeticone and simeticone in pharmaceutical formulation. J Pharm Biomed Anal 58:200–207

Moore JC (1964) Gel permeation chromatography. I. A new method for molecular weight distribution of high polymers. J Polym Sci A General Papers 2(2):835–843

Moore DE, Liu TX, Miao WG, Edwards A, Elliss R (2002) A RP-LC method with evaporative light scattering detection for the assay of simethicone in pharmaceutical formulations. J Pharm Biomed Anal 30:273–278

Otto DP, Vosloo HCM, Villiers MM de (2007) Application of Size Exclusion Chromatography in the Development and Characterization of Nanoparticulate Drug Delivery Systems. J Liq Chromatogr Rel Technol 30(17):2489–2514

Porath J, Flodin P (1959) Gel filtration: a method for desalting and group separation. Nature 13(183):1657–1659

Podzimek S (2011) Light scattering, size exclusion chromatography and asymmetric flow field flow fractionation: powerful tools for the characterization of polymers, proteins and nanoparticles. Wiley, New York

Pujolar JM, Marino IA, Milan M, Coppe A, Maes GE, Capoccioni F, Ciccotti E, Bervoets L, Covaci A, Belpaire C, Cramb G (2012) Surviving in a toxic world: transcriptomics and gene expression profiling in response to environmental pollution in the critically endangered European eel. BMC Genomics 13(507):1–22

Russ A, Berek D (2007) Enthalpy assisted size exclusion chromatography: part 2. Adsorption retention mechanism. J Sep Sci 30:1852–1859

Shan-Ying C, Chang-Qin H, Ming-Zhe X (2003) Chromatographic determination of high-molecular weight impurities in amoxicillin. J Pharm Biomed Anal 31:589–596

Striegel A, Yau WW, Kirkland JJ, Bly DD (2009) Modern size-exclusion liquid chromatography: practice of gel permeation and gel filtration chromatography, 2nd edn. Wiley, New Jersey

Subedi RK, Oh SY, Chun MK, Choi HK (2010) Recent advances in transdermal drug delivery. Arch Pharm Res 33(3):339–351

Suhre K, Meisinger C, Doring A, Altmaier E, Belcredi P, Gieger C, Chang D, Milburn MV, Gall WE, Weinberger KM, Mewek HW, Hrabe deAM, Wechmann HE, Kronenberg F, Adamski J, Illit T (2010) Metabolic footprint of diabetes: a multiplatform metabolomics study in an epidemiological setting. PLoS ONE 5(11):1–11

Sweetman SC (ed) (2009) Martindale 36, The Commplete Drug Reference. Pharmaceutical Press, London

Templeton DM, Ariese F, Cornelis R, Danielsson LG, Muntau H, Leeuwen HP von, Lobiński R (2000) Guidelines for terms related to chemical speciation and fractionating of elements. Definitions, structural aspects and methodological approaches (IUPAC Recommendations 2000). Pure App. Chem 72(8):1453–1470

Temyanko E, Russo PS, Ricks H (2001) Study of rodlike homopolypeptides by gel permeation chromatography with light scattering detection: validity of universal calibration and stiffnes assessment. Macromolecules 34:582–586

Torrado G, Garcia-Arieta A, De los Rios F, Menendez JC, Torrado S (1999) Quantitative determination of dimethicone in commercial tablets and capsules by Fourier transform infrared spectroscopy and antifoaming activity test. J Pharm Biomed Anal 19:285–292

Trathnigg B (2000) Size exclusion chromatography of polymers. In: Meyers RA (ed) Encyclopedia of analytical chemistry. Wiley, Chichester, pp 8008–8034

Walsh Z, Levkin PA, Jain V, Paull B, Svec F, Macka M (2010) Visible light initiated polymerization of styrenic monolithic stationary phases using 470 nm light emitting diode arrays. J Sep Sci 33:61–66

Williams A, Varela E, Meehan E, Tribe K (2002) Characterisation of nanoparticulate systems by hydrodynamic chromatography. Int J Pharm 242:295–299

WHO technical Report series 952, Evaluation of certain food additives, Sixty-ninth report of the Joint FAO/WHO Expert Committee on Food Additives. World Health Organisation, Geneva, 2009

Wu CS (ed) (1999) Column handbook for size exclusion chromatography. Academic Press, USA

Wu C (ed) (2004) Handbook of size exclusion chromatography and related techniques. Chromatographic science series, vol 91. Marcel Dekker Inc, New York

Yu Y, Mishra S, Song X, Lasanajak Y, Bradley KC, Tappert MM, Air GM, Steinhauer DA, Halder S, Cotmore S, Tattersall P, Agbandje-McKenna M, Cummings RD, Smith DF (2012) Functional glycomic analysis of human milk glycans reveals presence of virus receptors and embryonic stem cell biomarkers. J Biol Chem 287(53):44784–44799

ERRATUM

Inamuddin, A. Mohammad (eds.), *Green Chromatographic Techniques,*
ISBN 978-94-007-7734-7, DOI 10.1007/978-94-007-7735-4_2, pp. 19–54,

---

DOI 10.1007/978-94-007-7735-4_2

Chapter 2 This chapter is written by the following Authors:

Maria Tsyurupa
Zinaida Blinnikova
Vadim Davankov

Chapter 2 The chapter title is incorrect. It should read as "Ion Size Exclusion Chromatography on Hypercrosslinked Polystyrene Sorbents as a Green Technology of Separating Mineral Electrolytes".

Index

Inamuddin, A. Mohammad (eds.), *Green Chromatographic Techniques*,
DOI 10.1007/978-94-007-7735-4, © Springer Science+Business Media Dordrecht 2014

MIX
Papier aus verantwortungsvollen Quellen
Paper from responsible sources
FSC® C105338

If you have any concerns about our products,
you can contact us on
ProductSafety@springernature.com

In case Publisher is established outside the EU,
the EU authorized representative is:
Springer Nature Customer Service Center GmbH
Europaplatz 3, 69115 Heidelberg, Germany

Printed by Libri Plureos GmbH
in Hamburg, Germany